ROCK LIGHTHOUSES

of BRITAIN

CHRISTOPHER NICHOLSON

Whittles Publishing

Published by Whittles Publishing Limited

Dunbeath Mill, Dunbeath, Caithness, KW6 6EG, Scotland

www.whittlespublishing.com

ISBN 978-184995-137-1

Hardback edition ISBN 1-904445-27-6 © 2006 Christopher Nicholson

Second edition published 1995, reprinted 1999 (twice), 2000

First edition published 1983

There are some illustrations where, despite my best efforts, it has proved impossible to identify the copyright owners of some of them, or indeed the copyright may have lapsed. I trust this will not spoil the enjoyment of the illustrations themselves.

Printed and bound in Great Britain by Severn, Gloucester

CONTENTS

The present Eddystone
lighthouse and the stump
of Smeaton's tower under
heavy skies
(Author)

PREFACE TO THE 3RD EDITION

WHEN THE second edition of this book was published at the beginning of 1996, the automation of the British lighthouse service was in full swing. Indeed, the process that had been underway in earnest since the 1970s was, if anything, drawing to a close. Most of the major rock lights had lost their keepers and were controlled from a bank of monitors inside the Trinity House depot in Harwich, or the Northern Lighthouse Board headquarters in Edinburgh.

By the end of 1998 it was complete. The era of the manned British lighthouse had ended. There is now no such thing as a lighthouse keeper. All British lighthouses function entirely automatically – the silicon chip, cycle-charge generators, solar panels and telemetry have seen to that. They still require annual routine maintenance visits and fuel replenishment, but gone is the era when three men kept a twenty-four hour, seven days-a-week, fifty two weeks-a-year watch inside every lighthouse tower.

The final Trinity House rock station to lose its keepers was Hanois in the Channel Isles, a Nicholas Douglass tower very much in the traditional rock lighthouse 'design' – rising as it does from a sea washed reef. It also had the distinction of being the first rock lighthouse to be powered entirely by solar energy. The very last British rock lighthouse to be converted to automatic operation was Hyskeir in the Hebrides. The NLB finally withdrew the keepers here on 31st March 1997.

Above: The Skerries
(CJ Foulds)

Although, not strictly rock lighthouses, it is appropriate to mention here the final Northern Lighthouse Board and Trinity House stations to succumb to automation. When the lighthouse flag at Fair Isle South was finally lowered on 31st March 1998 in the presence of Her Royal Highness The Princess Royal it was the culmination of over 30 years of planning and conversion, and brought to an end a tradition of over 200 years of Scottish lighthouse keeping. The Chief Executive of the NLB James Taylor said, "This is truly a sad and memorable occasion, as we pay tribute to the men who have served the Board and the Mariner so well for over 200 years. From now on we provide the same service to the same high standards through an entirely automated network. The Lightkeepers, a truly remarkable and dedicated breed of men, are gone, but we will never forget them."

The last of all was Trinity House's North Foreland station on the Kent coast. This was the very last manned lighthouse in the United Kingdom, and at a ceremony attended by The Master of Trinity House, His Royal Highness The Duke of Edinburgh on 26th November

1998, the flag was lowered and the last six Trinity House keepers left the Service. The British lighthouse service was now totally automatic. The era had ended.

Since those two occasions the lighthouse service has entered a new century where, responsive as ever to the needs of the mariner, the two lighthouse authorities continue to provide a service as reliable and cost-effective as they always have. The NLB has had a particularly active programme of new lighthouse construction in response to local requirements. Not 'rock' lights in the mould of the examples in this book, but smaller, basic navigation aids known as 'minor' lights that function entirely automatically and require minimal maintenance. However, they have even built a light that would fit the description of a 'rock' lighthouse to improve the navigation through an extended deep water passage to the west of the Outer Hebrides. The uninhabited island of Haskeir has never had a navigational aid of any sort, but since October 1997 it has been the home to a 'major' light (that actually looks like a lighthouse) to aid vessels into and out of the oil terminals of Shetland.

With the pin-point accuracy of satellite navigation some lights have been deemed to be surplus to requirements and have had their upkeep and administration devolved to various pilotage and port authorities, or have been closed and sold as private dwellings. Spending a different sort of holiday in a former keepers cottage around the coasts of England, Wales and Scotland is now possible at a wide variety of locations.

Powering lighthouses with solar and even wind energy has become a common policy for both the NLB and Trinity House over many years and is clearly going to be the way forward for both authorities. Banks of solar panels are commonplace at both rock and shore stations – the energy they harvest is used to replace the diesel engines that used to recharge the batteries at each station. Power from light to produce a light in the dark.

SOUTH ELEVATION

Above: Samuel Wyatt's original Longships lighthouse of 1795. *(Trinity House)*

C.P.N.
May 2006

Left: The Smalls *(Steven Winter)*

ACKNOWLEDGEMENTS

I HAVE no doubt that were it not for the interest and enthusiasm of the many people I have made contact with in the course of researching and writing the many editions of this book, it could have so easily turned into a thankless task. Special mention must be made of Paul Ridgway, the public relations officer at Trinity House when I first started writing this book, who was so helpful in searching out information and photographs, and arranging for me to fly on various helicopter reliefs. Jane Wilson, Breda Wall, Howard Cooper and Vicki Gilson have provided subsequent and equally valuable assistance at Trinity House. At the Northern Lighthouse Board my thanks are due to Ian Archibald and Lorna Hunter for whom my persistent requests for information or photographs were never too much trouble.

I must also mention the many anonymous keepers, engineers and lighthouse board employees I have met on visits to their stations. Without exception they were only too willing to submit to my detailed questioning, and provided me with endless details and anecdotes about life in a rock station which space, unfortunately, does not permit me to recount in full. Mike 'Eddie' Matthews, then the principal keeper at St Ann's Head lighthouse – now the attendant at Lizard lighthouse, and Walter Boyd, the former bosun at Trinity House's Penzance depot, who made my helicopter visits to several towers such enjoyable and memorable occasions, must be mentioned in particular.

The success of these flights was due in no small part to the pilots and staff of Bond Helicopters Ltd., especially Mike and Geoff Bond, Dave Farendon and Dave Allen who were the pilots on the majority of my flights to Trinity House and NLB lighthouses.

There have been many people who have made contact since the first two editions of this book was published. I cannot name them all, but their kindness in doing so is equally valued. They have offered information or put me in touch with the source of other photographs that has enabled me to correct some details in the earlier editions, update the appendix tables, find alternative illustrations and add further information to each chapter, particularly with regard to automation. Jim Bain, a keeper on the Bell Rock and Eddie Dishon, an NLB engineer, have been generous enough to allow me to use their photographs, as have John Hellowell on the Isle of Man, and Steven Winter in Devon. Gerry Douglas-Sherwood, the archivist of the Association of Lighthouse Keepers, has managed to unearth some real treasures in their archives.

I have to mention here my regular correspondence with Arthur Lane, a former keeper at the Eddystone, whose articulate and amusing letters and emails have both educated and entertained me over the years since he bought the first edition in 1983. His knowledge of the Eddystone is encyclopaedic and his own work, *It Was Fun While It Lasted* – from the same publisher as this volume – is the most complete and hilarious account of the life of a lighthouse keeper I have ever read. Many of the interesting details about life in a rock station I have included come directly from Arthur's pen.

Lastly, but certainly not least, a special word of thanks to my wife, Sue, for her untold patience at being driven to many remote headlands while I strained my eyes at the horizon for a glimpse of one of the subjects of this book, or engaged a keeper in lengthy conversation at a lighthouse we 'mysteriously' arrived at during our travels. Taking a holiday whose itinerary no longer includes anything to do with lighthouses will make a welcome change for her!

C.P.N.

BRITAIN'S ROCK LIGHTHOUSES

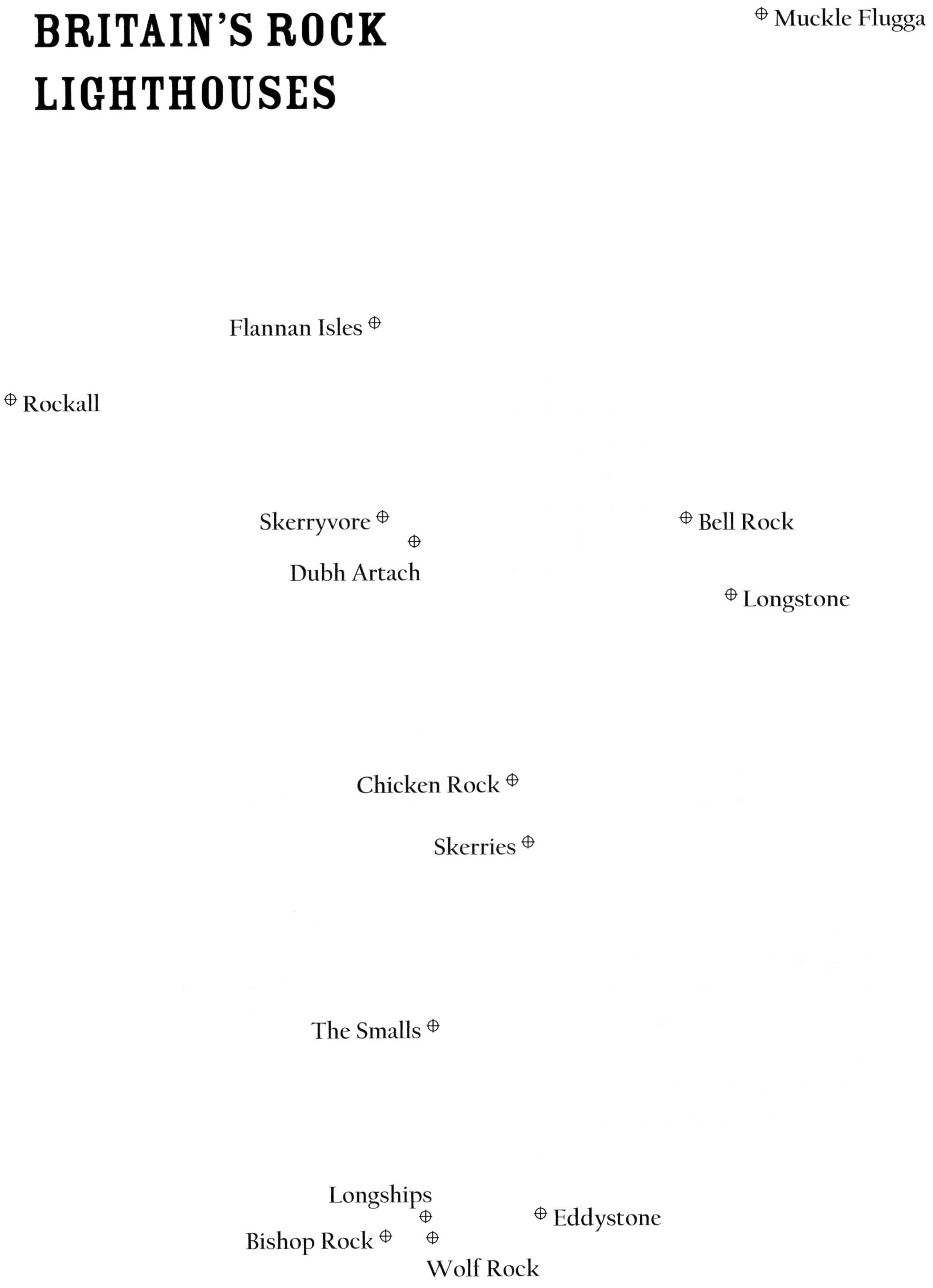

⊕ Muckle Flugga

Flannan Isles ⊕

⊕ Rockall

Skerryvore ⊕
⊕
Dubh Artach

⊕ Bell Rock

⊕ Longstone

Chicken Rock ⊕

Skerries ⊕

The Smalls ⊕

Longships
⊕
Bishop Rock ⊕   ⊕
Wolf Rock

⊕ Eddystone

INTRODUCTION

Right: Bell Rock – Britain's oldest surviving rock lighthouse. *(Royal Bank of Scotland)*

EVER SINCE Man has existed on the face of the earth he has been in continual conflict with the forces of nature, pitting himself against the elements in a vain struggle for supremacy. Such an association has shown that, for all his technology, Man is still woefully inadequate when involved in a head-on encounter with the elements in fury. Despite this, his persistent attempts to reduce such overwhelming odds are to his credit. In many fields these efforts have proved futile, yet there is indeed one area where he has worked to lessen the effects of such front-line encounters – to temper his relationship with what must be nature's most awesome force – the sea.

Man's challenge takes him everywhere upon the globe. Ever since the early days of maritime exploration he has learnt of the many moods of the sea; tranquil one minute yet a terrifying and destructive killer the next. Such a precarious existence, at the mercy of a force over which he had no control, was against all his natural inclinations. So it was that his innate sense of self-preservation brought about the first crude attempts to make his passage safe over the oceans of the world. A system was required which enabled mariners to venture across the seas, regardless of their state, night or day, without falling victims to the innumerable perils that lurked unseen. From these simple needs stem the whole of the world's 50,000 lighthouses and beacons, evolving through time from bonfires on cliff tops or beaches, to a complex system of lights, gongs, horns, bells, radar and satellite navigation that now provide a fail-safe system for the competent mariner. The hazards that once ensnared him are now clearly marked so that he may pass safely by.

Since their conception, lighthouses have stirred the emotions of the civilised world. There is something strangely romantic, if not mysterious, about an isolated tower of rock planted on a wave-swept reef sending out its beam of light across the turbulent waters below. Although by no means do I wish to belittle the land stations, their function is equally as vital as the rather more prestigious rock tower, yet the sight of a granite finger rising sheer from a half-submerged rock and subject to assaults by wind and wave of unprecedented ferocity, cannot fail to overawe.

The life of the keepers of these sentinels – when there were such things – was one of loneliness and isolation that inspired authors to take up their pens in tribute to these dedicated men. It was they who attended to the proper functioning of the light, regardless of the weather or its station. It was they who saw that shipping was warned of the unseen dangers below, both in clear visibility and in fog. It was to these anonymous men that many a mariner was indebted, for no career called for so much devotion to duty with so little obvious reward. With such a powerful combination of lighthouse and lightkeeper, locked together in the front line of the battle between Man and Nature, it is not surprising that literary works of both fact and fiction on this subject are numerous. Sadly there is no longer such an association. Fiction writers can, of course, still turn to these structures for tales of mystery and isolation, but the facts behind them can often surpass the writer's imagination for drama and excitement – something that I hope will emerge from this volume.

While there are a vast number of subjects I could have chosen, I intend to concern myself with a few of the more isolated or interesting examples of lighthouse engineering around the shores of Britain. It is these structures that invariably have the most spectacular histories. All the chapters that follow are based on fact. They are the tales of the dramas that occurred during the construction of these beacons and their subsequent histories. Some will be better known than others. Who, for instance, has not heard of the famous Eddystone lighthouses or Grace Darling's heroic deeds at the Longstone lighthouse, yet how many people know of the mysterious events that took place on the Flannan Isles late in 1900 – events that have still not satisfactorily been resolved to the present day?

Above: One of the original Trinity House plans for the Smalls lighthouse of 1861.
(Trinity House)

The chapters have been arranged in chronological order of the first structures on the different sites so that an idea may be gained about the history and technological developments in lighthouse engineering. However, I do not intend what follows to be a exhaustive study of the civil engineering or architectural styles of each light. Nor is it intended as a catalogue of the vessels that have come to grief on the various reefs. It is written for the many people who take an interest in lighthouses and enjoy tales of the sea in its ever changing moods. Above all it is a tribute to the courage and endeavour of the men who designed, built and manned these masterpieces, for such enterprise demands recognition.

Right: A Bell Rock keeper, dressed in collar and tie for the occasion, poses for a photograph taken from inside the lens of the lighthouse.
(Arbroath Museum)

1 OUT OF THE DARKNESS

a history of
the British Lighthouse service

THE coastline of England and Wales measures some 2,405 nautical miles, while that of Scotland, indented by its many sea lochs, extends to 6,214 miles. The process of lighting these coastal areas is one that has been continuing, albeit fitfully, since the Dark Ages. It is still not complete – the Northern Lighthouse Board has constructed several new lighthouses and beacons around the Scottish coasts over the past couple of decades to guide the massive vessels that travel to and from the Shetlands oil terminal through the treacherous waters around these islands.

Many of the early beacons were very crude affairs indeed – cliff top bonfires or coal-burning grates – totally inadequate by later standards, yet these first attempts to guide early mariners back to port or away from danger were the forerunners of the modern British lighthouse system. Unfortunately, while there were those who strove to lessen the danger of travelling the seas in darkness by maintaining such devices – notably the Church, shipowners or local councils – equally popular was the practice of erecting false beacons, luring ships to their peril and earning for the wreckers a lucrative income from the looting of their cargo. Since those early days, the building of lighthouses and beacons has kept pace with man's advances in technology until today we have an almost foolproof system to guide the mariners of the world safely around our coasts.

The administration of all the lighthouses, buoys and beacons of England, Wales and Scotland is the responsibility of just two organisations; the Corporation of Trinity House and the Commissioners of Northern Lighthouses. It is, perhaps, appropriate at this point to look briefly at the history of these two venerable bodies. In this way a better historical understanding will be gained of the individual lighthouses that follow.

By far the elder of the two bodies is Trinity House, which is empowered to construct and maintain the buoys, beacons and lighthouses around the coasts of England and Wales as well as the lighthouses of the Channel Islands and a solitary station in Gibraltar. Sadly, many details of the early history of Trinity House have been lost owing to the devastations of the Civil War, the Great Fire of London, a further fire in 1714, and the particularly disastrous enemy bombing of London in December 1940. However, it is probable that the organisation as we know it evolved from one of the seaman's guilds or associations which were in existence in the Middle Ages. These were essentially charitable organisations that took responsibility for retired seamen, their widows and orphans, as well as providing pilots and recommending suitable sites for buoys and beacons in their particular area. In the 16th century there were known to be such guilds in Dover, Hull, Scarborough, Newcastle, Leith and Dundee, while the most influential was probably the Trinity Guild based at Deptford Strond, an old mooring place on the banks of the River Thames. It had connections with the Catholic Church as its full and rather verbose title, 'The Master, Wardens and Assistants of the Guild, Fraternity or Brotherhood of the Most Glorious and Undividable Trinity and of Saint Clement in the Parish of Deptford Strond in the County of Kent,' confirms.

In 1514 King Henry VIII granted a Royal Charter to the Trinity Guild that allowed it to draw up rules for the benefit of English shipping and deal appropriately with any breaches of these regulations. When Henry was succeeded by Edward VI the Guild changed its name to the more familiar 'Corporation of Trinity House on Deptford Strond' to avoid drawing unwanted attention to their religious background, as it was at this time that the dissolution of the monasteries was in full swing. Similar charters to that of 1514 were also granted to the other guilds which also then became known as Trinity House.

The 1514 charter was confirmed by subsequent monarchs, Edward VI in 1547, Mary in 1553 and Elizabeth I in 1558, but it was not until 1566 that the first connection between Trinity

Opposite: Sunset over the Skerries lighthouse. *(C.J. Foulds)*

House and navigational aids was apparent. In that year an Act of Parliament was passed that granted the authority to, "…*the Mayster, Wardens and Assistaunts of the Trinitie House of Deptforde Strond, being a Company of the chiefest and mooste experte Maysters and Governours of Shipps…*" to, "*…from tyme to tyme, hereafter at theyre Wylles and Pleasures, and at their Costes make erecte and set up such and so many Beakons Markes and Signes for the Sea, in such Place or Places of the Sea Shore and Uplandes near the Sea Costes or Forelands of the Sea, onely for Sea Markes, as to them shall seeme most meete needeful and requisyte, whereby the Daungers maye be avoyded and escaped, and Shippes the better coome into their Portes without Peryll.*" Its authority was widened in 1594 by a further grant from Elizabeth I which allowed Trinity House to place buoys off-shore and in navigational channels.

Two points should be noted about these charters. Firstly, there is no mention in any of them about a structure called a 'lighthouse', only "beacons, marks and signs of the sea." This was to be a bone of contention between Trinity House and the Crown for many years to come as the Corporation argued that their Royal Charter implied that a lighthouse, which after all was definitely a 'sea mark,' was included within the scope of these documents and therefore to them had been granted the sole right to erect lighthouses around the coasts of England and Wales. Unfortunately, the law officers and law courts at that time disagreed, stating that such a right remained with the Crown. The second point of note is that there is no facility within the Royal Charters to allow for the expense of construction to be reimbursed, or for the collection of maintenance dues from passing shipping to be enforced. Such authority had to be applied for from the Crown after the completion of each project.

Any individual or collective body could erect a lighthouse in Britain provided they applied to the Crown for a patent. If granted, such a patent authorised its applicant to erect his lighthouse and collect dues for a specified number of years from shipping passing the light when they reached port. For this privilege, the patentee had to pay an annual rent to the Crown. The patentee could either be Trinity House itself or a private individual, but whoever applied for such a patent had to submit a petition to the Crown detailing the need for such a structure from local seamen and merchants, who also had to pledge their willingness to pay an appropriate levy when the lighthouse was operational. Each application was carefully scrutinised by the Privy Council, which often consulted Trinity House when private applications were received, when such matters as the proper management of the proposed lighthouse was considered, as well as the social standing and financial viability of the individuals concerned. If all was found to be in order then the patent was granted, although Trinity House were somewhat jealous of their Royal Charter and objected to many applications using what became their standard argument that if a lighthouse was necessary at a certain location then they themselves would have erected one. This was somewhat difficult to reconcile as they also frequently repeated a statement, to the effect that they did not propose lighthouses until they received petitions from local interests for one. The outcome of this somewhat muddled situation was that Trinity House often received the patent for a lighthouse and then granted a lease for a private individual to build it for them. This was an eminently satisfactory arrangement for Trinity House as the entire expense of erection was born by the lessee while the Corporation received rent if the construction was successful.

The problem of collecting light dues was overcome by most patentees by entering into an arrangement with the Trinity House agent based at the London Custom House. He in turn employed agents in about one hundred ports who collected the dues from shipping and were paid a percentage commission for doing so. This figure could be as much as 20-25 per cent in the period up to the mid-19th century and led to light due collectors in some of the busier ports becoming very rich men indeed, with considerable influence in the affairs of Trinity House. The rates of lighthouse dues were also very steep. The exact figure varied with the size of the vessel, its cargo and destination, and also how many lights it would pass en route. Foreign vessels were always charged double. Such high rates were necessary, argued Trinity House, to enable them to fulfil their original commitment to look after the welfare of retired or distressed mariners and their families. The owners of private lighthouses had no such drain on their income, yet the rates they charged were equally excessive. It is not difficult to see the enormous financial advantages gained from the granting of a patent for a private lighthouse that resulted in the alleged advice to prospective patentees to "...watch the moment when the King is in a good temper to ask him for a lighthouse."

Parliamentary Select Committees of 1820, 1834 and 1842 recommended reductions in the percentage commission paid to light due collectors and that a standard rate of dues should be levied. The rate of commission was duly slashed to a mere 3¾ per cent, and a further Select Committee of 1845 said that "...if Light-Dues shall continue, they ought to be collected by the officers of Customs in every port of the United Kingdom free of all charge whatsoever...". This was duly authorised and since then light dues have been collected by HM Customs and paid to

Left: A golden ray of light strikes the elegant façade of Trinity House on Tower Hill, the building from which all the lighthouses of England and Wales are administered.
(Mark Denton Photographic)

the Board of Trade, now known as the Department of Trade and Industry. The auditing of accounts pertaining to the General Lighthouse Fund is done by Trinity House.

The patent of each lighthouse had to be renewed at intervals when the efficiency and reliability of the light were again considered by the Privy Council who could grant a further patent to its original builder or return the light to the jurisdiction of Trinity House. In the early days of lighthouse construction there was a very real risk that the cost of maintenance and the keepers' wages, together with the rental paid to the Crown, could exceed the dues collected from vessels passing the light. However, after about the middle of the 18th century when there was a rapid growth in sea trade, particularly with the New World, many lighthouses which were run by private individuals became lucrative sources of income for their fortunate owners, a fact which had not gone unnoticed by Parliament and Trinity House.

An Act of Parliament of 1822 decreed that no further extensions to lighthouse patents should be granted and that Trinity House should be at liberty to purchase private lighthouses from their patentee if both were agreeable and a suitable sum could be negotiated. Naturally enough, there were many patentees who were unwilling to part with their own private 'goldmines' and stubbornly resisted the coercions of Trinity House. They would have been less than pleased therefore when a Parliamentary Committee of Enquiry in 1834 made detailed investigations into the management and profitability of all British lighthouses that required their owners to produce the accounts for the year of 1832. With much reluctance the

Above: Trinity House's crest bears the latin inscription 'Trinitas In Unitate' – 'Three In One' – a reference to its original functions of controlling the lighthouses and lightships, providing pilotage in major ports, and as a charitable organisation for distressed mariners.
(Trinity House)

figures were eventually produced and revealed, as was expected, soaring profits from most of the private lights.

All this, however, was soon to come to an end. As a result of the Parliamentary Enquiry an Act of 1836 was passed giving Trinity House the authority to purchase the patents of all the lighthouses of England and Wales which were still the property of private individuals. There were only ten of these yet their purchase cost a staggering £1,182,546. Exact details of several of these transactions appear in later chapters. The sale of the last private lighthouse was completed by 1841.

The 1836 Act was also a milestone in the history of Trinity House for two further reasons. Since they were now responsible for the management and maintenance of all English and Welsh lighthouses, a standard rate of duty could be charged, regardless of how many lights were passed in the voyage, which was a vast improvement on the old system whereby different lighthouses levied duties at varying rates, all of which had to be calculated and collected for each light passed. Also, all light dues collected by the Corporation were henceforward used solely for maintaining and improving their lighthouses, rather than being diverted for charitable purposes in helping retired seamen and their dependants – the original purpose of the Trinity House Guild. Since 1836 Trinity House has been able to erect a lighthouse in England or Wales without first obtaining a patent from the Crown.

The day to day running of Trinity House is the responsibility of ten 'Elder Brethren' who swear solemn oaths of loyalty to the sovereign. In addition they are advised by several hundred seamen known as 'Younger Brethren.' The Corporation also employs numerous scientific, executive and clerical officers. The headquarters has moved from Deptford Strond across the river, and is today found at 'Trinity House' on Tower Hill. From this magnificent

Right: The Northern Lighthouse Board's imposing Georgian headquarters are in George Street, Edinburgh. Over the front door is a model of the Bell Rock lighthouse. *(Northern Lighthouse Board)*

building the administration of a service which is now responsible for 72 lighthouses, 10 light vessels/ light floats, 412 buoys, 19 beacons, 48 radar beacons, and 7 satellite navigation reference stations is conducted. Because these stations are spread along the length of the English and Welsh coastline, the Channel Islands and Gibraltar, Trinity House originally had six depots, but these have been rationalised to just two – at Harwich and Swansea. Their two sea-going tenders, *Patricia* and *Mermaid*, operate from these bases that also have the facilities to maintain and repair lights and buoys. Harwich is also the base from which all of the Trinity Houses navigational aids are monitored 24 hours a day. *Alert* is a smaller 'emergency response vessel' based in the Straits of Dover that can deal with wrecks or buoy collisions within a matters of hours. Around the coastline of England and Wales there are also many local and harbour authorities which have been granted the authority to establish and maintain their own navigation lights and buoys that are subject to regular inspections by Trinity House officers who are required to be consulted before any alterations are enforced.

The lighthouses of Scotland and the Isle of Man are administered and maintained by the Commissioners of Northern Lighthouses, often referred to as the Northern Lighthouse Board. By comparison with Trinity House, their history is a straightforward affair with few of the legal intricacies which are so characteristic of their English counterpart.

Prior to the mid-18th century Scottish foreign trade was almost negligible and therefore the need for lighthouses of no great concern. In 1760 only six existed, but with the upturn in maritime trade during the latter half of the century demand grew for lighting various sites, particularly along the estuary of the River Clyde. In due course a body known as the Cumbrae Lighthouse Trustees was inaugurated in May 1757 to erect a lighthouse on Little Cumbrae, an island straddling the mouth of the Clyde estuary. The Trust was authorised by the Parliament of the day and was moderately successful in its aims. This led to an Act of Parliament being passed in 1786 to establish four more lighthouses, '...for the security of navigation and the fishermen in the Northern Parts of Great Britain' away from the Clyde, and for the trustees of these lights to collect appropriate dues from passing vessels when complete. The Trustees would be: the Lord Advocate and the Solicitor General for Scotland, the Lord Provosts of Edinburgh, Glasgow and Aberdeen, the Provost of Inverness and the Chairman of Argyll and Bute, the Sheriffs-Principal of the counties with a coastline, plus various magistrates. They would henceforth be known as the Commissioners of Northern Lighthouses and the composition of this body today remains exactly the same as in 1786. Its headquarters are in George Street, Edinburgh.

The Northern Lighthouse Board inherited the responsibility for the Manx lighthouses as a result of an Act of Parliament in 1815. At that time the merchants of Liverpool and the Isle of Man were petitioning Trinity House to construct lights on the island. The levy Trinity House proposed for doing so was considered so extortionately high – justified by the Corporation once again as being required to allow them to fulfil their charitable obligations – that the merchants demanded the lights be administered by the Commissioners of Northern Lighthouses whose dues would therefore be lower. This arrangement exists to the present day.

The Northern Lighthouse Board is a much smaller organisation than Trinity House. It controls and maintains 212 lighthouses, 154 buoys, 47 beacons, 25 radar beacons and 4 satellite navigation reference stations. Although it functions as an independent body it is interesting to note that any new lighthouse construction in Scottish waters requires the consent of Trinity House before it can take place. Both lighthouse services come under the general control of the Department of Trade and Industry.

The cost of the lighthouse services provided by Trinity House and the Northern Lighthouse Board are met from a common fund known as the General Lighthouse Fund (GLF) whose principal income is 'light dues' levied on different classes of shipping calling at ports in the UK. Until April 1993 these were collected by HM Customs and Excise, but since that date they have been the responsibility of the Institute of Chartered Shipbrokers in all areas except the Isle of Man where Light Dues are still collected by Customs. The Secretary of State for Transport has responsibility for the GLF and sets the level of dues to be charged in order to maintain the Fund in proportion to costs of the service. Because of the programmes of automation and conversion to solar power it has actually been possible to reduce light dues by an astonishing 40% over the last decade. Both Trinity House and the Northern Lighthouse Board are non-profit making bodies that are funded by their users at no cost to the taxpayer.

Like most self-financing bodies these days, ways to improve the efficiency of the service the give and reduce overheads are always high on the agenda. Nowhere is this more apparent than in the British lighthouse service. Both Trinity House and the NLB have, in recent years, undergone major reviews in this respect with the result that both bodies are now significantly smaller in terms of manpower, the numbers of navigational aids under their control, and the size of their supporting fleets, than they were just a decade ago. Rapidly advancing technology has meant that any lighthouse can now be made to function automatically without the need for keepers – with tremendous financial savings.

Trinity House announced the findings of a major three-year 'Navaid Review' during 1987 that proposed the discontinuing or transferring to local harbour and port authorities of almost one third of the navigational aids it was previously responsible for. The recommendation that 63 of these aids to navigation were no longer necessary reflected technological advances both within the lighthouse service and on board vessels themselves. The development and improvement of the 'racon' (radar beacon) has also meant that that certain major floating aids such as lightvessels could be replaced by buoys or 'lightfloats' equipped with racons.

Another joint review by all three UK General Lighthouse Authorities published in 2005 was intended to address current and future requirements for national and international shipping around our coasts up until 2010. The idea of these regular reviews is prudent management and planning for the future – "to ensure that the aids to navigation provided by the GLAs in the interest of general navigation are cost-effective and continue to meet the present changing need of all mariners and comply wherever possible with internationally-accepted criteria and that timely reviews of the GLAs' aids to navigation are carried out to facilitate financial, operational and engineering planning."

Above: The NLB crest carries the inscription 'In Salutem Omnium' – 'For the Safety of All' (*Northern Lighthouse Board*)

This particular review recommended a whole raft of different measures from discontinuing certain lights while establishing new ones, altering the character of their lights, reducing or increasing their range, replacing lights with racons, discontinuing their fog signals, or handing them over to local harbour authorities. Almost every combination of change you can imagine will take place at various locations around our coasts in the next few years. There will then be another review to reassess the situation as navigation requirements will no doubt change further in line with developing technology.

It is true to say that both lighthouse services recognise that less reliance is now placed on traditional aids to navigation by many vessels equipped with well maintained on-board navigational aids such as satellite navigation that can be accurate to within five metres. However, there is also a significant number of vessels using the waters around our coasts that are not so well equipped or their on-board aids are liable to frequent malfunction or to unskilled use by inadequately trained personnel. It is for the latter vessels, and to reduce the risk to life and the possibility of environmental pollution from an incident, that the majority of our lighthouses, lightvessels, buoys and beacons need to be maintained.

Opposite: Sunset at Bell Rock – seen as an inverted image through the lens in the lantern. (*Jim Bain*)

In this way the British lighthouse service of the 21st century will probably be a completely automatic yet highly sophisticated network of lighthouses, lightvessels, buoys, fog signals, radar beacons and satellite navigation stations that provide just as reliable a service to the mariner as it has always done – at a fraction of the cost.

2 EDDYSTONE

the beginning

IT IS perhaps fitting that the first example in this volume of notable lighthouses is the one generally regarded as the most famous lighthouse in the world. Some 14 miles south-south-west from Plymouth and 9 miles south of Rame Head, Cornwall, rise three ridges of viciously hard red gneiss rock which have been broken and shattered by a ceaseless assault from the sea. The three fingers of rock are only visible above the water for half a mile as 23 jagged and sloping pinnacles, converging in an arrowhead pointing landwards. Their extent below sea level, as they climb steeply from the bed of the English Channel, is considerably longer. At high spring tides a mere 3 ft of the reef remain uncovered, and more often than not these are shrouded with foam and spray which can be driven upwards to spectacular heights. The formation of the ridges and the complex tidal peculiarities of the Channel stir the water around these rocks into a troubled, seething swirl, even on the calmest of days. In rough weather conditions they are positively evil. The notoriety of this group has long been known; the vicious tidal rips through the narrow channels and the turmoil of currents between the rocky teeth were responsible for the early seafarers bestowing upon this reef a name befitting these phenomena. They called it 'eddy-stones', and it is amongst these rocks that the dramas of the building and destruction of the famous Eddystone lighthouses have taken place.

The history of this morbid cluster and the beacons placed upon them is a truly remarkable legend amongst lighthouse stories and a tribute to the tremendous determination, courage and daring of those men who laboured under indescribable conditions to warn others of the peril on which they toiled. During the 17th century world trade was on the up-turn, more and bigger ships were being constructed, and sea ports once again began to thrive. The city of Plymouth, England's westernmost port of any size, was no exception. In that century alone the city had doubled its population, the naval dockyard had been established there, and the port was prospering from trade with the New World. It possessed a superb, sheltered harbour and circumstances were altogether favourable for its continued expansion – apart, that is, from one thing. Just 14 miles from its extensive waterfront was the Eddystone reef, to overshadow any other assets of which the city could boast. Lying as it did across the approaches to Plymouth Sound the ships it swallowed were without number. So feared was this reef it was not uncommon for ships to be wrecked on the rocks of the Channel Islands or the rugged coastline of northern France, such was the fanatical desire of some captains to give these terrible rocks as wide a berth as possible.

It was obviously imperative that something be done to stem this wholesale destruction of life and property, so in 1664 a petition was presented to Trinity House for a lighthouse to be erected on the Eddystone. It was dismissed by the Elder Brethren as desirable yet impractical, an understandable reaction in the circumstances as nowhere on the globe was a structure to be found which could provide a precedent to work from.

Eddystone continued to claim its victims for a further 30 years until 1694 when a patent was granted by the Crown 'to erect a Lighthouse or Beacon with a light upon the rock called Eddystone off Plymouth … as safe direction for ships hereafter to avoid that dangerous Rock upon which the lives of so many of our good Subjects have perished …'.

It was not until the following year that Trinity House, who were obviously still sceptical about the practicality of such an undertaking, arranged for a local man, Walter Whitfield, to build the tower, 'at his own cost and entire financial risk'. For his trouble he was to receive all the dues collected from shipping passing the light for the first five years after its completion, half the dues for the next 50 years, whereafter Trinity House would collect all such monies. Even in these early days it would appear that there was someone at Trinity House with a shrewd business sense, for if Whitfield failed they lost nothing, but the erection of a durable

Opposite: Too close for comfort? Not really. This dramatic view of Smeaton's stump and its successor was taken by a diver who surfaced in one of the channels of the reef. The storm damage to the top of the stump is especially noticeable in this view, as is the ring of new solar panels around the outside of the lantern. *(Dan Bolt)*

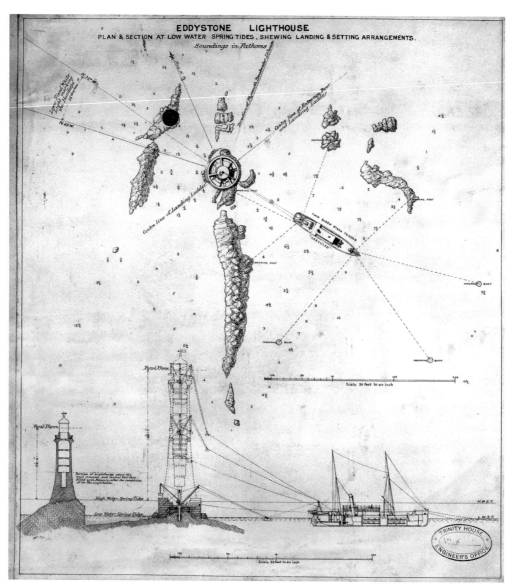

Left: The original Trinity House plans of the Eddystone reef itself, showing the sites of the Smeaton and Douglass towers, and the mooring arrangements for the *Hercules* as she unloads stones for the new lighthouse.
(Trinity House)

structure meant a considerable financial income in the future. It is thought that Whitfield then visited the reef to make his plans but was so daunted by what he saw and what he would have to produce, that he realised he would be incapable of completing such a mammoth task.

On to the scene now comes the most colourful character in the whole of the lengthy Eddystone saga. His name was Henry Winstanley and he was indeed a remarkable man. Born in 1644, Winstanley lived in Saffron Walden, Essex, and amongst his many talents could list inventor, showman, designer, competent conjuror, engraver and businessman; but above all was his fame as a practical joker. His house was filled with all manner of weird and wonderful devices which were intended to puzzle and amuse, ranging from chairs which changed shape or imprisoned the occupants, and apparitions that appeared from the floorboards, to mirrors which made the apparent position of a doorway alter, thus causing the unfortunate victim to walk into a wall when wishing to leave. To the casual observer life in the Winstanley household must have been difficult to take at all seriously, yet here was a man who, though unqualified as an architect or civil engineer, was about to become involved in setting a lighthouse on one of the most bleak and exposed stations around our coasts – and with no existing example to work from. The howls of derision could be heard all the way from Saffron Walden to Plymouth Hoe.

Winstanley was also a shipowner. From the capital he had accrued as a result of his other business interests he had bought five vessels. In August 1695 one of these was lost on the Eddystone, and four months later a second, the *Constant*, followed the fate of her sister ship. With almost half his complement of vessels lost within six months Winstanley was quite rightly furious. Arriving at Plymouth he demanded to know what was being done about this menace. The position regarding a lighthouse was explained to him, and also the lack of someone to build it. True to form, Henry Winstanley, jack of all trades, promptly offered to build the good people of Plymouth their lighthouse, although even he must have pondered over the wisdom of his gesture after realising the Herculean nature of the task. To build a lighthouse on a wave-swept rock 9 miles from the nearest land was a daunting challenge indeed which would take a man of tremendous foresight and courage if it was to be accomplished. The whole of Plymouth prayed

that Henry Winstanley, self-confessed showman and practical joker, would be just such a man who could rid them of this slur on their civic pride. Agreements were signed which gave Winstanley the entire profits for the first five years, after which he would share them equally with Trinity House. First, though, he had to build his lighthouse.

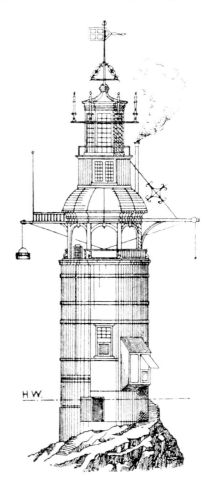

In 1696, at the age of 51, Winstanley set about his work. If completed successfully it would surely be the crowning glory to his varied career; if he failed all credibility in the future would be lost. His first task was to decide exactly where to build his light. Along the length of the three main ridges which compose Eddystone there was not one single area of level rock. All the ridges rose from the sea on their western flanks at an angle of about 30 degrees, to fall vertically on their eastern faces. Such a formation caused an almost permanent wall of spray to be thrown up as the waves broke over the crest of the ridge or else met the vertical face head on. Only on the westernmost ridge was there sufficient rock left above high water capable of supporting the base of a tower. This is where Winstanley chose to build.

Left: This is Winstanley's second attempt at an Eddystone lighthouse after the strengthening in 1699 of his first tower. Even this wasn't strong enough to resist the 'Great Storm' of 1703.

Plans were drawn up during the previous winter; plans which involved the fixing of 12 iron bars, 3½ ins in diameter, in a circular pattern on the rock. These would then form the framework for a solid masonry base. By late June 1696 work was commenced, but it was soon apparent that the rust coloured gneiss of Eddystone was reluctant to be pierced. Armed with picks, Winstanley's men laboured throughout the summer at just this one task. The gneiss was of an incredible hardness and blunted their tools rapidly. Spare picks were brought along but these too suffered the same fate. As if this were not enough, the rock was being continuously washed by breakers, the men were cascaded with salt spray, and achieving a firm foothold on the wet, seaweed covered rock while swinging a pick was far from easy. At the end of the day, of which as little as two hours could actually have been spent on the rock, the men were faced with an exhausting journey back to Plymouth. Indeed, it was not so much the tedious nature of the work that caused Winstanley to despair during this first season, but more the journey to and from the rock, a strength-sapping row which could take anything up to 9 hours should wind and tide be against them. Often, after leaving Plymouth, the wind would suddenly rise or change direction making a landing on the rock impossible and disheartening the men for the weary return journey.

Eventually, after many months of erratic toil on the rock, hampered by "out-winds and the in-rush of the ground-swell from the main ocean", the 12 holes were at last ready to receive their iron stanchions. It was now late in October and Winstanley knew that the winter gales would soon be upon him. Time was scarce. He was particularly anxious to fix these bars before the weather made it impossible and he would have to finish work for the year. The iron was shipped out in several journeys, each bar being placed in the hole and the space around it filled with molten lead, which when solidified held the bar firmly in position. It must have been with some relief that Winstanley left the reef, with the tide rising rapidly, on an October day in 1696. Pulling through the channels in the reef, 12 stout iron bars could be seen breaking the skyline. He prayed they would still be there when he returned the following year to continue with his labours.

During 1697 work progressed steadily. Winstanley returned to find the iron bars still in place and set to forming the base around them. Local Plymouth stone was used, but this proved no easy task to unload from an open boat heaving in what could be a considerable swell. With the iron rods acting as a framework the stone blocks were cemented around them until a solid pillar, 14 ft in diameter and 12 ft high, rose from the sloping rock. Once a flat surface became established for the men to work on progress became rapid.

It was during this season that the now famous tale of Winstanley's kidnapping took place. England was engaged in one of the numerous wars with France which were commonplace during the history of the two nations. The Admiralty were naturally keen that nothing should hinder Winstanley with his humanitarian efforts and so provided a guard ship, the *Terrible*, to keep an eye on things at the reef. It was a dull and tedious life on board with precious little to occupy the men, so it was not surprising that the Captain of the ship, Timothy Bridge, decided to abandon his post in favour of a nearby French merchant ship which would provide rich pickings and relieve the monotony for the crew. During the chase, one of the notorious Channel fogs descended, preventing the capture of the French ship and also the return of the *Terrible* to Eddystone. In the meantime, the reef had been approached by a French privateer which despatched a party of armed men who captured the unfortunate Winstanley and set his workers adrift in an open boat. Winstanley was carried off to France and brought before Louis XIV who was somewhat perturbed at the incident. The officer responsible was punished, Winstanley was showered with gifts and returned to England, with the alleged comment from Louis that, "Your work is for the benefit of all nations using the seas. I am at war with England, not with humanity."

Despite this set-back, and the valuable time which had been lost because of it, by the end of the season the solid base of the tower was complete. Winstanley could now spend this winter in an easier frame of mind than he spent the last, as the hardest part of his task was behind him. Over the winter months a revision was made to the original plans. In the following season the first job was to increase the base diameter by 2 ft and the height by a further 6 ft to produce an even more stable base on which to plant the superstructure. The pace of construction was brisk. Above the base came the store room, and above this was the elegantly named living or 'state' room. As soon as the walls of the living accommodation were high enough, Winstanley hoped to be able to use them as a permanent base where he and his men could live, thus eliminating the tiresome return journey to Plymouth every day.

By June of that year the hollow section was ready for occupation, but the baptism of his structure turned out not to be what Winstanley would have wished for. He describes the first night in the tower in his account of the construction:

Being all finished, with the lantern, and all the rooms that were in it, we ventured to lodge there soon after midsummer for greater dispatch of this work, but the first night the weather came bad and so continued, that it was 11 days before any boat could come near us again, and not being acquainted with the height of the sea rising, we were nearly all ye time neer drowned with wet and all our provisions in as bad a condition, though we worked night and day as much as possible to make shelter for ourselves. In this storm we lost some of our materials although we did what we could to save them. But the boat then returning we all left the house to be refreshed on shore, and as soon as the weather did permit, we returned and finished all...

The final adornments were soon shipped out and, by November 1698, the 80 ft tower was ready to be lit. On 14th November Henry Winstanley lit for the first time the tallow candles suspended in the lantern gallery, with the comment that, "It is finished, and it will stand forever as one of the world's most artistic pieces of work." This was the signal for great rejoicing in Plymouth and surrounding areas. People rushed to the cliff tops and strained to catch a glimpse of the yellow glow reported by fishermen far out to sea. There is no doubt that Henry Winstanley became a public hero, he had done what no one had attempted before and he had achieved what many had thought impossible. His tower would doubtless end the many centuries of death and destruction caused by this wicked reef, yet still there were those who scorned. Looking at this structure it was not difficult to see why, for no building looked quite so misplaced on a reef in the open sea as did Winstanley's tower. It was not so much the shape of the structure which amazed the sceptics, more the excessive amounts of decoration bestowed upon it. External ladders, carved balcony railings, ornate eaves, and a massive wrought iron wind vane were all to be seen. It really was a most unusual lighthouse, but it was there, warning ships off the Eddystone reef and away from a watery grave. This was reason enough for the townsfolk of Plymouth to revel. In the midst of all this jubilation there was still one person absent from the celebrations – Winstanley himself. The explanation is again to be found in his own narrative:

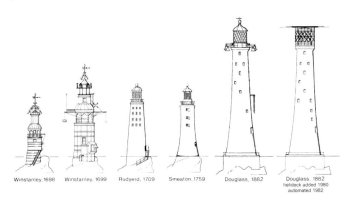

Winstanley, 1698   Winstanley, 1699   Rudyerd, 1709   Smeaton, 1759   Douglass, 1882   Douglass, 1882 helideck added 1980 automated 1982

Left: An interesting comparison of the five different lighthouses that have been placed on Eddystone reef, with the two different versions of James Douglass's tower. *(Trinity House)*

We put up the light on the 14th November, 1698. Which being so late in the year it was three days before Christmas before we had a relief to get ashore again, and we were almost at the last extremity for want of provision, but by good providence then two boats came with provisions and the family that was to take care of the light, and so ended the year's work.

The weather during that winter was not at all good. Frequent storms assailed the tower at regular intervals, so Winstanley was particularly keen to inspect the light at the earliest opportunity. He made his first visit in the early spring and was not happy with what he found. The constant wave and spray action had prevented the cement between the blocks of the solid base setting properly. It was now in desperate need of re-pointing if it was to survive another winter like the last. Also, the keepers related horrific tales of how the tower shook and shuddered when hit by waves which frequently obliterated the light from the lantern, over 35 ft above the sea. A drastic situation called for drastic measures. Winstanley decided that he would completely encase this existing tower with a newer and stronger one. The base diameter was increased by 8 ft to 24 ft, and the height by 2 ft to 20 ft. The joints between courses of blocks were covered by bands of iron stretched around the whole circumference of the tower in an attempt to lessen the sea action on the cement. The upper parts of the structure were increased by the same proportions and now rose to a magnificent 60 ft from its base to the lantern gallery. The external ornamentation was, if anything, increased. A balcony allowed the lantern windows to be cleaned from the outside, numerous cranes adorned the building for unloading supplies, a flagpole and several candlesticks were affixed to the exterior of the lantern room, multilingual inscriptions were carved into many of the outside walls, and the whole edifice was once again topped by an ornate wind vane. On many occasions Winstanley's second light, which was lit during 1699 and cost him £8,000, has been likened to an oriental pagoda. Once again the sceptics heaped derision upon it, being in their opinion even more ridiculous than his first effort. How could such a structure possibly perform the function of a lighthouse when it had obviously come straight from a Chinese cemetery?

Right: An early engraving of Rudyerds lighthouse, surrounded by an extraordinary number of galleons, showing the tapering oak planking to good effect.
(Devon Library Service)

Nevertheless, Winstanley's second tower survived its first winter and the next two after that. The pale yellow glow from its 60 candles warned of the evil on which it stood, and throughout this time not one single vessel was lost on the Eddystone. Every year the light became the subject of morbid curiosity as to how long it would remain on the reef. The people of Plymouth were well aware of the many projections from the tower, which it was said the sea would have no difficulty in tearing at and thus toppling the whole structure from its base. Yet still it remained. Winstanley, wishing to silence his critics, was once heard to remark that so solid was his lighthouse that he could wish for nothing better than to be in it during, "the greatest storm that ever was".

Unfortunately, he was to get his wish. He was in his tower on the evening of 26th November 1703 effecting repairs to the fabric of the building when one of the most devastating storms the British Isles has ever experienced occurred. Coming as the climax to a fortnight of steadily worsening gales, it was likened to a tornado as it swept across the southern counties in an orgy of death and destruction. Thousands of people died in the few short hours the storm was at its peak, whole houses were swept away like dust, chimney pots were hurled through roofs, rivers burst their banks flooding acres of land, and ships sheltering from the hurricane were driven against one another or on to the shore where they were reduced to kindling with the reported loss of 8,000 sailors. Over 17,000 mature trees were dragged up by their roots in Kent alone and flung through the air as if they were matchsticks, pedestrians were lifted bodily for many feet, and church steeples were sent crashing downwards. The utter devastation of such a cataclysmic force cannot be imagined. It became known quite simply as the Great Storm. A contemporary historian wrote of this tempest:

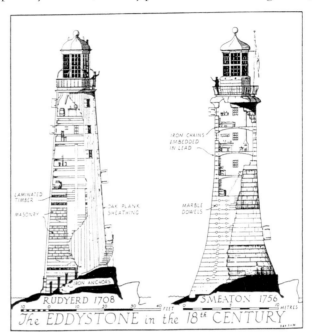

Left: A comparison of the two towers that stood on the Eddystone in the 18th century.
(from a drawing by Douglas Hague)

The effects of the great wind were fearful in the extreme, trees fell like grass before a great scythe, leaden roofs of churches and cathedrals were rolled up like grey carpets and carried from sight, while ships strong enough, with gallant crews, which managed to struggle into Plymouth, reported the light of the Eddystone still shining bright and people ashore fell to their knees to thank the Almighty.

Out on the Eddystone the seas were mountainous. The light from the candles was seen up until about midnight when the sheets of driving rain and decreasing visibility blotted it from view. When the sun rose on the following morning and the fury of the storm had abated, there was no sign that any structure had ever existed on the Eddystone rocks, apart from the twisted remains of 12 iron piles which had once held the base to the reef. It was the end of the first chapter in the history of the Eddystone lighthouses, but more significantly, it was the end of Henry Winstanley, philanthropist and lighthouse builder, who had died fulfilling his greatest wish.

Despite the many criticisms that were levelled at Winstanley, he was a resolute and determined man, constantly changing and amending his plans in order that he might achieve what he believed to be the most appropriate structure for the reef, all financed from his own pocket. He had proved that it was possible to place a beacon on a sea-washed rock which gave sufficient illumination to warn passing vessels of its presence. The fact that it stood for over three years without the loss of a ship is testament to this. Indeed it would probably have existed for a great deal longer had it not been for the undreamt severity of the Great Storm.

If we are to look for a reason to explain the failure of his light it was, in all probability, nothing to do with the excess of decorations, for these would offer little surface area, and therefore resistance, to wind or waves. The weakness was more likely to be in the base and its method of attachment to the rock. It would appear from Winstanley's own plans that this relied solely on the 12 iron bars which later proved inadequate under real stress. Only the bottom half of his tower was stone, the upper rooms were of wooden construction and so added little to the weight and wave-resistant properties of the light.

Left: John Smeaton (1724 – 1792), the Yorkshireman who built the fourth Eddystone lighthouse, which can be seen in the background of this engraving.

Whatever its shortcomings, there was one fact that became overlooked during the general aftermath of events. Eccentric though he was, Henry Winstanley had placed the first lighthouse on a wave-swept rock anywhere in the world. That fact above all should not be forgotten. Just two days after the sea had claimed Winstanley and his tower a merchant ship, the *Winchelsea*, was bound for Plymouth with a cargo of tobacco and 67 passengers from America. Out in the Atlantic they had battled through heaving seas and mile-long breakers, the likes of which none of the crew had seen before, but now in the final approaches to their destination the captain peered eagerly for the flickering glow of the Eddystone light. Visibility was poor so the whole crew lined the decks and strained their eyes to pick up the light they knew must be there. When roaring water and foaming white surf were spotted by a lookout it was too late. Everyone realised what was happening but were powerless to prevent it. The helpless vessel ploughed on to the westernmost ridge where its bottom was ripped out. Only two men escaped with their lives, the rest perished on the Eddystone reef, a reef which until two nights before had boasted a lighthouse and no wrecks for over five years. All that had now changed, and the loss of the *Winchelsea* only highlighted the need for a new light to check the Eddystone's insatiable greed for ships.

Although Winstanley had demonstrated the feasibility of putting a light on the Eddystone, Trinity House were still reluctant to exercise their right to erect a similar structure. Instead, in 1705, a 99-year lease was granted to a Captain John Lovett for the putting up of a light on the rock in return for an annual rent of £100. For its construction Lovett commissioned the services of a silk mercer from Ludgate Hill, London, called John Rudyerd, who again appears to have had little or no experience in the fields of architecture or civil engineering, a prerequisite one would imagine for the dubious business of erecting a lighthouse on a rock in the open sea. He was born into filth and degradation, the son of a Cornish labourer, but ran away to London on his twelfth birthday where he prospered in the silk trade. Nevertheless, Rudyerd produced a structure which was every bit as brilliant as one that might have been designed by a trained man in this field. His design departed radically from Winstanley's in that his tower was composed mainly of wood, because, Rudyerd said, only wood could give the necessary flexibility to resist the force of the waves. A detailed study of the previous two lights showed up the weaknesses of Winstanley's bold efforts – "This was the work of an artist, not an engineer" – so a completely new approach was adopted by Rudyerd. He produced plans for a slim, tapering, conical structure, sheathed in ships' timbers and devoid of any projections but for an external ladder to the entrance door and railings around the lantern gallery.

In July 1706 Rudyerd commenced his work. The same site that Winstanley had used was chosen again, presumably for much the same reasons. Also, the same difficulties of breaking the rock encountered by the previous engineer still faced Rudyerd and his gang. It was his intention to work the sloping surface of the rock into a series of steps on which the base of the structure could be placed, but so hard was the gneiss that the resulting efforts were very rudimentary indeed. Into these crude steps were cut a series of 36 holes to form the method of attachment for the tower to the rock. Rudyerd rightly considered this to be one of the major weaknesses of Winstanley's design. Each hole had a dovetailed section. They were all approximately 16 ins deep with a width across the top of 7 ins, narrowing to 6 ins halfway down, before widening to 8 ins at the bottom. These holes were intended for a series of iron bars or 'branches', 6 ft high with two branches in each hole, being of such a shape that when placed in the hole it would be impossible to remove them together. This, Rudyerd hoped, would be the key to the stability of the whole structure. He reasoned that any water left in these holes when the bars were set in position could lead to corrosion, and therefore a weakness in the final structure. He overcame this problem by an instance of reasoned brilliance that characterised the whole of Rudyerd's work on the reef. Each hole was dried with a sponge and immediately filled with tallow, preventing any more water from entering. When the branches were ready for placement they were heated strongly and plunged into the hole, thus melting the tallow. The remaining spaces around the bars were then filled with molten lead or pewter which would displace any remaining tallow and solidify to form a water-free foundation. It was a simple but nevertheless ingenious process which worked exceptionally well on a reef that was continuously running with water. When these structures were examined by the person who was next to have the task of erecting a beacon on Eddystone they were found to be in the same condition as if they had just left the blacksmith's forge, such was the effectiveness of this rock and metal union.

As Rudyerd's design was to rely heavily on wood he enlisted the services of two experts in this medium, Messrs Smith and Norcutt, master shipwrights of Her Majesty's Naval Dockyard at Woolwich. With their expertise in jointing and shaping timber, and his foresight, Rudyerd had planned to produce a structure which could withstand the worst that Eddystone could offer. Their first task, now the iron branches were set in position, was to build up a platform of wood by alternating rows of heavy oak blocks on the stepped rock bed so that this base rose above the height of the surrounding reef. These timbers were bolted to the iron branches and also to each other, forming a solid foundation 10 ft high and 23 ft across, and adhering closely to the rock.

For the construction of the actual tower Rudyerd used another unique method. It consisted of alternate layers of timber and granite, five rows of granite blocks, two of timber and so on, the rows of

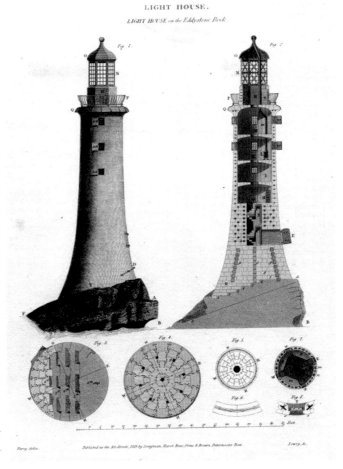

LIGHT HOUSE.

LIGHT HOUSE on the Eddystone Rock.

Left: John Smeaton based his design on the trunk of an oak tree – a design which subsequently set a precedent for all lighthouse engineers to come. The way in which the blocks of granite were secured to one another using dovetailed joints, joggles (square marble blocks) and trenails can be seen in this engraving of the construction details.

stone decreasing as the height increased. The whole was bound together by a complex series of clamps, wooden nails, iron bars and bolts to form a coherent mass. This sandwich type of construction was totally solid for 9 ft above the highest point of the rock, at which point a central hollow section was introduced to accommodate the staircase and passage from the entrance door. In this central well was a large, circular wooden pole, akin to a ship's mast and planted firmly into the rock. This was the core of the structure to which the various layers of timber or granite could be bolted for even greater stability and resistance to the lateral force of the waves. At the top of the stair well the foundation was now 36 ft high and contained 273 tons of granite.

Above this was built the accommodation, extending upwards for a further 34 ft and consisting of four rooms; a store room, living room, bedroom, and below the lantern, a kitchen. To this tower Rudyerd added his own particular trade mark. The whole of the foundations and walls of the living quarters were encased by 71 upright planks of oak, jointed with the precision only a craftsman could obtain, and tapering gently to form a smooth outer skin to Rudyerd's innovative tower. The joints were caulked and the entire outer surface liberally coated with pitch to resist saturation by the waves. The tower now stood 71 ft high and tapered from almost 23 ft at its base to just over 14 ft at the top. The light was housed in a wooden lantern house and consisted of 24 tallow candles. Although the tower was not finally complete, Rudyerd lit the candles on 28th July 1708. What a magnificent structure now stood on the Eddystone reef, in complete contrast to its predecessor. Only two years in the building, Rudyerd's tower owed much to maritime traditions and techniques; the central wooden mast, the oak planking sealed with pitch, and the whole edifice being under the supervision of two shipwrights from a naval dockyard. Throughout its construction Rudyerd showed tremendous planning and meticulous attention to detail which made certain his structure was to stand on the reef for some considerable time to come.

Unfortunately, he saw very little of his brainchild, for in 1713, only five years after its inauguration, he died. By 1723, during a routine inspection of the tower, it was noticed that sea worms had attacked and eaten into the lower portions of many of the planks in the outer skin, particularly those which were regularly immersed by the tides or drenched with spray. John Holland, a shipwright in the naval dockyard at Plymouth, was despatched to effect repairs. At first he merely renewed sections of the planking but when, during later years, it was apparent that this was going to be a recurring ailment of the tower, Holland experimented with covering the base of the planks with copper sheets. This appeared to make little difference, and the lessees resigned themselves to the fact that a continuous programme of renewal and replacement would be necessary if the tower was to continue to be serviceable. To aid him in his work Holland drew up detailed plans of the outside of the structure on which he could mark exactly which planks had been replaced and when. He died in 1752 when his work became the responsibility of his protege, Josais Jessop.

Rudyerd's tower continued to warn ships off the Eddystone for 46 years until one fateful night in December 1755. That evening the wind was biting, and in the tower as usual were the three keepers, the oldest of whom was Henry Hall, an incredible 94 years old yet, "of good constitution and active for his years". He was on duty during the small hours of 2nd December when, during a routine inspection of the lantern room he found it to be on fire. Immediately he opened the trap door into the room the fire, which was probably only smouldering until then, burst into flames which were fanned by the freshening wind outside. Unable to waken his companions, thought to be the worse for drink, Henry Hall attempted to fight the fire single-handed. With only a leather bucket and a tub of rain water on the lantern gallery, he gallantly flung the water upwards into the lead cupola which conducted the candle smoke out through the vent in the roof. It was here that the fire probably started due to a spark from a cracked chimney pipe leading from the kitchen stove in the room below, through the lantern room and out through the roof. The lead covering the candles was encrusted with the soot and grease from nearly half a century's candles and so readily smouldered when ignited.

Hall's companions finally staggered up the stairs into the acrid black smoke, but by then it was too late. The fire had got a grip and was roaring through the dry timber, causing the lantern room roof to collapse. While hurling buckets of water upwards a bullet of molten lead dripped from the remains of the roof towards the upward gazing Hall who emitted a blood-curdling scream, "God help me, I'm on fire inside!" He claimed it had passed down his throat. The others were clearly sceptical and, anyway, there were more important matters to attend to than Hall, who apart from difficulty in communicating, appeared to be under no great suffering after his initial cries.

Left: An early photo of Smeaton's tower taken sometime between 1845 and 1882 because there is clearly a Fresnel lens in the lantern. It's a busy day at the lighthouse with at least one man in a cradle painting the outside, two on the lantern gallery (one hoisting a flag and the other using a telescope) and a ghostly figure down on the rock itself.

All three men were quickly driven downwards through the rooms by the searing heat and choking fumes until they eventually had to take refuge on the rock itself. They huddled in a tiny crack under the external ladder which offered slight protection from the rain of red hot bolts, molten metal and burning wood which cascaded down around the cowering keepers.

In the meantime, the inferno had been spotted from the mainland by a Mr Edwards, "a man of some fortune and more humanity," and a boat despatched to try to effect a rescue. It was not until 10 o'clock in the morning, 8 hours since the fire first started, that an open boat picked its way into the reef to rescue the frightened, weary and sea-drenched keepers, who had to be hauled though the icy water on the end of a rope to reach the boat. Henry Hall was still complaining about the metal he had swallowed. Little of the tower now remained, the solid base was well aglow and it was obvious to the boatful of onlookers that Rudyerd's tower was beyond redemption. The Eddystone reef had lost another lighthouse.

Arriving at Plymouth one of the keepers immediately took flight and was never heard of again, so greatly had his ordeal affected him. Henry Hall was put under the medical supervision of Dr Henry Spry but still mumbled about his awful experiences and the molten lead. Local opinion put it down to the shock of that night to a person of his advancing years. To be badly burnt, choked by suffocating fumes and then have to spend the night crouched on the reef dodging flaming missiles before being dragged to safety through chilling waters, had obviously turned his mind. Hall made a slight recovery before dying 12 days after he was brought ashore. A post-mortem revealed a flat lump of lead, 4 ins by 1¾ ins wide and of 7 ounces 5 drams weight, occupying most of his stomach. After all his supposedly demented ravings, Henry Hall had been telling the truth and had survived for 12 days with a quantity of once molten metal in the pit of his stomach. This very lump of lead can be seen today in the Royal Scottish Museum, Edinburgh. Dr Spry wrote an account of this case and presented it to the Royal Society. Few of the Fellows of that day gave his paper much credibility, an attitude which incensed Spry. For the sake of his reputation he performed experiments with dogs and chickens and proved that they could live after having molten lead poured down their throats!

As the last few smouldering timbers of Rudyerd's light fell into the turbulent waters around Eddystone there was once again a feeling of despondency in Plymouth. He had produced a tower, the likes of which had never been seen before, which had stood for almost half a century against the worst of Atlantic storms with only some of the planking needing replacement. Now, the very thing that made the tower unique was the very thing that caused its destruction. Had it been built of stone the flames might not have spread at such an alarming rate and there might have been a chance of extinguishing them, or at least rebuilding the wooden part.

Nothing at all remained on the reef except the heat-buckled iron straps so meticulously planted by Rudyerd. Again, no ships were lost on the Eddystone while the beacon stood, so the necessity for its replacement was beyond question. Indeed, it was not long before steps were in hand for a fourth lighthouse, leaving the name of John Rudyerd to join that of Henry Winstanley in the list of gallant men who had temporarily tamed Eddystone.

During the life of Rudyerd's tower, the ownership of Eddystone had changed hands. Since 1724 the chief shareholder was Robert Weston, an astute and caring man who saw the need for a replacement light as a matter of priority. Determined to have a structure that would outlast the previous three, he contacted the Royal Society, the leading scientific body of the day, to ask their advice as to its constructor. Without hesitation they at once recommended one man who, as a result of his efforts on the Eddystone, was to change the whole science of pharology in years to come. His name was John Smeaton.

Smeaton was a Yorkshireman, born in 1724, the son of a solicitor. He was a mathematical instrument maker by trade but throughout his childhood showed a far deeper interest in things mechanical. On the strength of several papers he had read to the Royal Society, he was elected a Fellow of that noble institution at the unprecedented age of 28. Instrument making soon became unprofitable so Smeaton turned his talents to the engineering field he excelled in and loved so much. He was contacted by Weston and after due consideration accepted the challenge of Eddystone, commenting that, "The time has come to build another light to adequately illuminate this very dangerous reef which is a constant threat to seamen." Would he be the man who could finally tame this malevolent rock?

In February 1756 Weston and Smeaton met in London to finalise details. Smeaton was a shrewd man who at once obtained an agreement that he should have a free hand in the design of the new light, and be able to incorporate any features he felt necessary, without intervention from Weston and his fellow shareholders. Smeaton knew they were in favour of a duplicate wooden structure in the image of Rudyerd's. With all to his satisfaction he left to prepare plans. What emerged from Smeaton's fertile mind was subsequently to set a precedent for lighthouse design throughout the world. Study of Winstanley's and Rudyerd's towers led him to the conclusion that his two main problems would be to make sure that the structure was heavy enough to resist any displacement by the waves, and also strong enough to prevent it vibrating during times of storm. To overcome these problems he had taken a bold decision early in his planning, namely, to build his tower entirely out of stone. This would afford the necessary weight while at the same time remaining rigid. However, where Smeaton's true genius showed was in the shape of the tower. Rather than repeating Rudyerd's design of a sloping cylinder, he enlarged the diameter of the base and made the tower sides slope inwards in a concave

Opposite: One of the original hand-painted Trinity House plans dated January 1878 for its new Eddystone lighthouse.
(Trinity House)

EDDYSTONE PROPOSED NEW LIGHTHOUSE

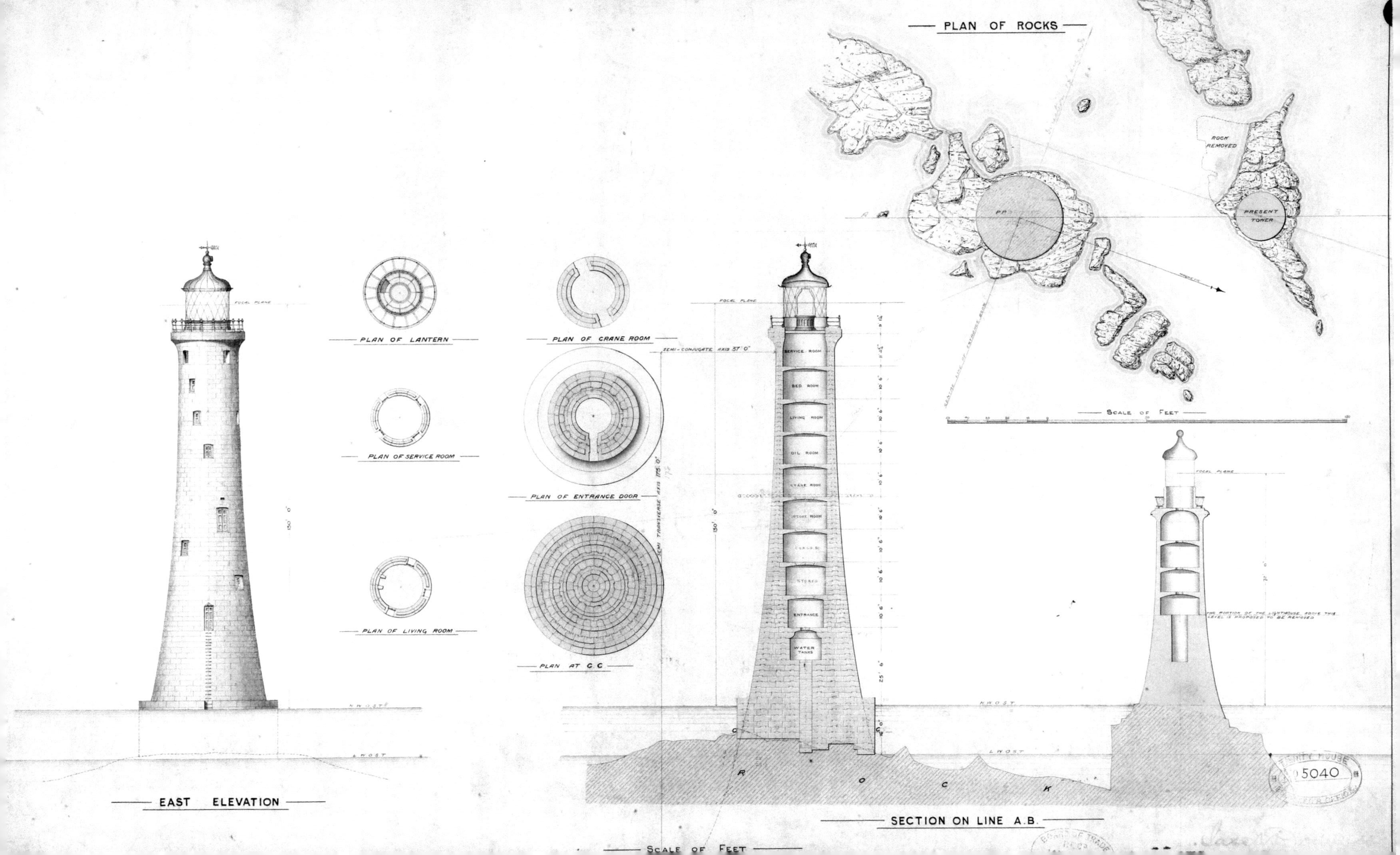
PLAN OF ROCKS
ROCK REMOVED
PRESENT TOWER
SCALE OF FEET
FOCAL PLANE
PLAN OF LANTERN
PLAN OF CRANE ROOM
PLAN OF SERVICE ROOM
PLAN OF ENTRANCE DOOR
PLAN OF LIVING ROOM
PLAN AT C C
SEMI-CONJUGATE AXIS 37' 0"
SEMI TRANSVERSE AXIS 175' 0"
130' 0"
SERVICE ROOM
BED ROOM
LIVING ROOM
OIL ROOM
CRANE ROOM
STORE ROOM
STORES
ENTRANCE
WATER TANKS
H.W.O.S.T.
L.W.O.S.T.
R O C K
THE PORTION OF THE LIGHTHOUSE ABOVE THIS LEVEL IS PROPOSED TO BE REMOVED
EAST ELEVATION
SECTION ON LINE A.B.
SCALE OF FEET
5040

curve, gradually becoming nearer to parallel as the height increased. To all intents and purposes he had based this design on the gracefully tapering outline of an oak tree, and explained the reasoning behind it as follows:

...as for my plans, put it this way – the English oak tree withstands the most violent weather conditions; so I visualise a new tower shaped like an oak. Why? Because the oak tree resists similar elemental pressures to those which wrecked the lighthouse [presumably he means Winstanley's]; an oak tree is broad at its base, curves inwards at its waist, becomes narrower towards the top. We seldom hear of a mature oak being uprooted!

He continued in more technical terms:

Connected with its roots, which lie hid below the ground, it rises from the surface therefore with a large swelling base, which at the height of one diameter is generally reduced by an elegant curve, concave to the eye, to a diameter less by at least one-third, and sometimes to half, of its original base. From thence, its taper diminishing more slowly, its sides by degrees come into a perpendicular, and for some height form a cylinder.

This was a design of rare perception, a design which was to make the name of Smeaton a legend amongst future lighthouse engineers. To strengthen the structure still further Smeaton decided that each stone block in every single course would be dovetailed to its neighbour with a joint similar to that used in carpentry, securely binding all the blocks to one another and making the whole mass solid. Unfortunately, this did not overcome the problem of securing each layer of blocks to those above and below it. With another novel idea, the engineer decided to use heavy marble 'joggles' (marble plugs 1 ft square), which were sunk into various blocks of each course so that half the joggle projected above the upper surface and fitted into a corresponding hollow in the course above. Nine joggles were used in each course, one in the centre and eight arranged in a circular pattern around it. Not only this, but every single block had two oak 'trenails' driven through it into the course below which, together with the marble joggles, would bind the whole structure into a solid, coherent monolith of stone, strong enough to resist the very worst Atlantic storm.

Weston and his fellow shareholders were enthusiastic. The Admiralty and Trinity House accepted his plans. All that remained for Smeaton was actually to visit the rock; all his detailed planning had been done without a visit to the site that was to occupy his attention for much of the next few years. To Smeaton, Eddystone was just a name on a sea chart as he had never set foot in the West Country in his life. He arrived in Plymouth on 27th March 1756 and quickly sought out Josais Jessop, the shipwright who was in charge of the running repairs to the planking of the previous tower before its demise. Jessop knew all the intricate details of the reef and structure of Rudyerd's tower so Smeaton found him an invaluable contact, to the extent that he was subsequently to take him on as his assistant.

During his two months stay in Plymouth, Smeaton was only able to land on the reef four times out of ten visits. The rest of the time was spent in choosing stone, finding a base for operations with a waterfront, taking on men, and generally finalising details before work could commence in earnest. It was during these two months that Smeaton learnt much through personal experience of the difficulties in landing on the reef and of the exhausting journey to and from Plymouth. Many a time he set sail but was unable to land, having instead to be content with gazing at the curtains of spray shrouding the rock. As an outcome of these many abortive and wearying journeys Smeaton decided a floating workshop and living quarters were a necessity if they were to snatch every available second on the rock and complete the undertaking in a reasonable time.

On one of his rare landings on Eddystone, he was able to carry out practical tests and observations on the behaviour of the sea, the hardness of the rock, and the best methods of piercing it. He also made detailed measurements so that a scale model could be built on dry land. Smeaton was thorough in his working and forward planning to the extent that he spared no detail to become conversant with its intricacies in the few precious hours he had on the rock.

The men he employed were divided into two groups of 12 plus a foreman, who would take it in turns to work equal periods on the rock or in the base at Plymouth, the gang out on Eddystone getting a higher wage than those on shore. Smeaton was a fair and just employer who drew up a *'Plan for carrying on the works and Management of the Workmen'* which was a detailed guide to procedures, rules and wages for his men, who correspondingly worked without undue fuss or strike throughout the whole of the time it took to complete the tower.

Work began on the reef on 3rd August 1756, the target for this short working season being to cut steps into the rock to take the dovetailed blocks for a solid base. Winstanley and Rudyerd had both experienced difficulty in chipping into the gneiss so Smeaton's work, requiring greater precision, would be that much more difficult. With intermittent spells of good weather the work proceeded apace. The system of shifts for the two gangs of men worked well; those out on the reef lodged between tides in the *Neptune*, an 80-ton herring boat, bought originally to provide a floating light adjacent to the rocks. Because of legal difficulties with Trinity House she was never used for this purpose but served admirably for the floating workshop and barrack.

With the steps eventually complete, the working season closed on Monday 22nd November. It was a typical winter's day with grey skies and a strong gale blowing. Smeaton and his gang were faced with the daunting task of navigating the rather awkward *Neptune* back to Plymouth when, without warning, the wind freshened further and the luckless crew found themselves in the grip of a near hurricane, their sails ripped to tatters while being propelled rapidly westwards by the ever-strengthening gale. They were well to the west of Land's End and heading for the Bay of Biscay before the storm abated sufficiently for them to change course and return. After leaving Eddystone on Monday afternoon they finally anchored in Plymouth Sound on Friday morning. This was a salutary reminder for Smeaton that no matter how well he thought he had planned, the sea was going to resist him to the bitter end.

He was not idle during the winter months, far from it. Huge blocks of Cornish moorstone, hard resistant rock for the outer walls, and the softer Portland stone for the interior, were waiting to be cut, shaped and dovetailed with absolute precision. A wooden template was made of each block in case of loss. Smeaton himself experimented with quick-drying cement in an attempt to lessen the chances of waves dislodging the blocks before the mortar had set.

Early in the morning of Sunday 12th June 1757 the first block of Smeaton's lighthouse was hoisted from a pitching boat on to the reef by a pair of wooden sheerlegs, block and tackle, and windlass. Engraved with the date '1757', it weighed 2½ tons and was to take two tides that day before it was finally secured by chains, wedges, mortar and trenails. The other three stones of the first course were laid the following day. Owing to bad weather on the 15th when the sea tossed five blocks from the reef, it was not until 30th June that the 13 stones of the second course were in place.

The 25 stones of the third course, and the 33 of the fourth course were secured during July, and by 11th August the sixth course stood proud of the reef, bringing the tower level with the top of the rock, and 8 ft above the heaving seas below. Instead of laying part courses on a stepped foundation, Smeaton and his men would now be laying complete, circular courses of stones. They must have been somewhat relieved at achieving the hardest part of their task with such comparative ease. The weather had been kind to them and serious delays were few. By the end of that season three further courses were intact and four more precious feet above the grasp of the waves had been gained.

At this point Smeaton made a curious journey to the headquarters of Trinity House in London, to ask if the Elder Brethren would rather he complete the rest of the tower in wood. They declined his offer and instructed him to continue with his original plans.

The third season on the reef started on 2nd July 1758. On 8th August the fourteenth course, 12 ft above the crest of the rock, was finished – so too was the solid base of the tower. Everything from now onwards would have a hollow section, either the passage from the entrance door and central stairwell, or the actual rooms of the light. By late September the lowest room, a store, was all but complete when Smeaton broke for the winter. He had intended to place some kind of light on the half-completed tower a year ahead of schedule, but petty-minded bureaucracy from Trinity House prevented this.

The gangs started the final season on 5th July 1759 with renewed vigour, for they could see that the end was in sight. After three years of toil and misery, living in the squalor of the *Neptune*, their goal was nearly complete which gave them renewed enthusiasm for their work. The remaining three rooms rose with great speed. Each was 12¼ ft in diameter with walls 2 ft 2 ins thick. Sixteen blocks were required to form one complete course of stones and all were secured to those above and below by metal cramps and the marble plugs. The second room was complete by the 21st, the third by the 29th, and the final 46th course was finished soon afterwards. At the top of the tower a stone cornice was laid around its circumference to deflect any rogue waves climbing the sides of the pillar away from the lantern glass. The masonry was at last complete on 16th August 1759 and, much to Smeaton's satisfaction, was only a mere eighth of an inch out of true in a total height of 70 ft. This accuracy was a vivid example of the skill and craftsmanship required from the masons whose standards never fell below such exacting tolerances during the whole of the construction.

Left: A really dramatic shot of the 'undercut' that caused the replacement of Smeaton's tower on the Eddystone. The entrance door has been 'bricked up' and the gun metal dog steps have been continued to the top of the stump.
(ALK archives)

The lantern was now installed, containing tackle for raising and lowering a chandelier carrying 24 candles, which, "will cost precisely 1/6½d an hour to burn". The keepers took up residence on 8th October while the finishing touches were being put, one of these was an inscription '24th August 1759 Laus Deo' on the final stone laid. Smeaton left the tower on 9th October to "…anchor in Plymouth Harbour with a flowing tide to the great joy and satisfaction of all concerned." The fourth Eddystone lighthouse was lit by the three keepers on Tuesday 16th October 1759 although, as if by fate, Smeaton was prevented from being there because of rough seas.

At last, after three gruelling years of conflict between Man and Nature, the Eddystone reef was once again conquered. It had taken the genius of a Yorkshire engineer, 1,493 blocks of stone weighing almost 1,000 tons, 700 marble joggles, 1,800 oaken trenails and £40,000 to do it. Out of the three years from start to finish, as little as 16 weeks were actually spent on the reef.

It was a splendid effort and an inspiring sight for all who gazed upon Smeaton's triumph. He made intermittent visits during the following years to inspect the light, yet found it just as it had been completed. On one of these visits he noted that while still 7 miles from the lighthouse the glow from it appeared, "...very strong and bright to the naked eye, much like a star of the fourth magnitude."

Smeaton's design was to change the whole course of lighthouse engineering in the future. His graceful, sloping lines were to be copied by almost every eminent designer to come, earning for himself the title 'Father of modern lighthouses'. His tower stood proudly on the Eddystone reef, defying wind and waves, for a time far in excess of any of its predecessors. Its fame became such that it featured on the English penny, at intervals, right up until 1970 and the Southern Railway even named one of their most powerful main-line steam locomotives No 34028 *Eddystone* when it emerged from their Brighton workshops in 1945 – the only British steam engine ever to take the name of a lighthouse. The whole of Britain soon knew of the achievement of this brilliant engineer, yet curiously Smeaton did not capitalise on his fame. He only designed one other lighthouse, at Spurn Head on his native Yorkshire coast. This is not the end of the Eddystone story. The sea was not to be beaten, although for over a century it was held in check.

During the long span that Smeaton's tower stood on Eddystone, important changes were taking place in the field of pharology. Most significant was the development by a French engineer name Argand of a new oil lamp which incorporated a circular wick covered by a cylindrical glass tube. It was capable of a brilliance undreamt of by earlier engineers and was therefore much used in the new lighthouses of the day, or for replacing the candles in existing structures. Eddystone's tallow candles were finally superseded in 1811 by 24 Argand lamps, costing Trinity House £3,000, the Elder Brethren having since taken over the expired lease on the light in 1807. In 1845 a Fresnel lens system was installed in the lantern using a four-wick lamp. This gave a beam of light visible for over 13 miles.

Shortly after the conversion to Argand lamps, the lighthouse was visited on several occasions by Robert Stevenson, Chief Engineer to the Northern Lighthouse Board in Edinburgh. On one of these visits signs became apparent that all might not be well with the tower. Smeaton had heard tell how his structure shook and vibrated when assailed by heavy seas, and these incidents were recounted by Stevenson on one of his landings. Surely nothing could be seriously wrong as the tower was still standing, apparently without detriment. Later structural examinations revealed that the actual tower was reasonably sound, yet the rock on which it stood was

Left: On 19th August 1879, the Duke of Edinburgh visited Eddystone to lay the first stone on the reef. This view was obviously taken from Smeaton's tower and shows the coffer dam around the workings and the steam-ship *Hercules* in the background. *(Devon Library Service)*

slowly being eaten into on one side by the sea, whose ceaseless pounding had hollowed out a portion of rock along the line of a fault. Robert Stevenson, himself a much respected engineer, was one of the first to sow the seeds of doubt about this alarming find. He commented thus: "It is shaken all through, and dips at a considerable angle...and being undermined for several feet it has rather an alarming appearance...were I connected with the charge of this highly important building...I should not feel very easy in my mind for its safety."

The situation was obviously serious. Nearly 1,000 tons of granite were resting on a piece of rock that was slowly but surely being eaten away by the sea, causing alarming vibrations to pass up the tower. Smeaton himself had noticed this cavity in its infancy, and it was perhaps with some apprehension that he wrote, many years previously, "Whether this cavity in the rock may ever proved of any detriment to the building is not to be determined with certainty."

Trinity House were in an almost impossible situation, for it was the rock, and not Smeaton's construction, that was giving rise to their fears. Had it been a weakness in the stones or their joints then perhaps a cure could have been effected. This was not the case, it was the Eddystone reef itself that was crumbling and Trinity House were powerless to stop it. The year was 1877, and after strengthening operations in 1837 and 1865 things on the Eddystone were no better. The impact of a heavy wave breaking against the tower was still transmitted through the stone-work in a frightening manner. It had stood for nearly 120 years on that lonely rock and few people now doubted that the end must surely be in sight. At a British Association meeting of that year, the Chief Engineer of Trinity House, James Douglass, announced that surveys had been made of the reef and existing tower which confirmed the deterioration in both. He found the tower itself in, "…a fair state of efficiency; but, unfortunately, the portion of the gneiss rock on which it is founded has been seriously shaken by the incessant heavy sea strokes on the tower, and the rock is considerably undermined at its base…" A new lighthouse was to be built and Douglass was already working on the plans. This was the end. After years of speculation a decision had finally been taken. Smeaton's tower was "a masterpiece of engineering" that had served faithfully for over a century, but in the end the sea had won. Unable to destroy the tower it had done the next best thing, and now its days were numbered. The whole of Plymouth grieved the loss of yet another lighthouse on the Eddystone.

Right: Block by block Douglass' tower on the Eddystone creeps upwards. In this 1879 picture there are approximately 20 masons posing for the camera during an exceptionally low tide and placid sea. *(Trinity House)*

James Douglass came to the Eddystone with his reputation already made. This would not be a reef on which he would have to prove his worth as did Winstanley, Rudyerd and Smeaton, for in Douglass' wake were three magnificent towers on the Bishop Rock, Wolf Rock, and the Smalls, all of which are dealt with later in this volume. In 1846 he designed the characteristic helically framed lantern, many of which still exist on today's lighthouses. His prowess and undoubted skill shone like the beacons he built, so raising another wave-swept tower on the Eddystone should prove little problem to an engineer of his standing.

His design for the Eddystone followed very closely that on the Bishop Rock. It still retained Smeaton's classic, curving outline, but this time the tower was set, not on the reef itself, but on a solid cylinder of masonry, 44 ft in diameter and 22 ft high. This new design was based on sound engineering principles; the cylinder of stone planted on the reef would break the lateral force of the waves as they crashed against it, reducing the chance of any rogue breakers climbing the sides of the light as their main energies would have been spent against the base of the pillar. It was a design that had proved itself elsewhere on sites every bit as bleak as the Eddystone, so Douglass could see no reason why he should not employ the same design there.

Before work could be commenced a major decision had to be taken. The sea was undermining the western ridge where all previous lights had stood, thus eliminating this site for the new work. A fresh area altogether was required. Douglass chose the central of the three ridges, at a point some 40 yards south-south-east of Smeaton's tower. Unlike the old position, this was submerged at high tide, requiring a huge, circular coffer dam 7 ft high and 7 ft thick to be raised around the chosen area before work could start. Prior to every session the stone and concrete barrier had

Right: Some of the masons responsible for the incredible accuracy of Douglass' tower pose on the final few courses of the lighthouse prior to their despatch to the reef in 1881. The 'dovetailing' of the blocks can be clearly seen. The third figure from the right standing on the top course is none other than James Douglass himself who has come to watch the final block of the final course being dry set in the yard of the De Lank quarry, Wadebridge.
(Trinity House)

to be pumped dry before the rock could even be seen, and every tide filled it up again, submerging their efforts until the next time that wind and waves permitted them on to the reef.

Douglass and his men, assisted by his son William, started the arduous task of setting another tower on the reef on 17th July 1878. It was now the Age of Steam and he was considerably aided in his task by a propeller-driven steam tug, the *Hercules*, which served to transport men and materials from the workyard at Oreston on the River Laira, out to the reef. Under favourable conditions it was possible to reduce his predecessors' weary 6-hour slog to something a little under 2 hours. This made them far more independent of the weather and so a longer working season was possible, extending on occasions from February to December. In that first year work on the coffer dam continued until almost Christmas, when a quarter of the dam was complete. It took over 40 landings to get just this far. The foundations of the dam were below the level of the tides, allowing only 3 hours' work at a time, much of this being spent up to the waist in water, attached to a lifeline, or clinging to an iron post when an unexpectedly large wave broke over the workings. Even so, 1,500 cu ft of rock were removed by the end of the season.

A contemporary work had the following to say about conditions on the reef:

Left: A rare photo taken early in 1882 showing one of the two huge fog bells being hoisted to the top of the tower where it will be hung from one of the two projections seen above the last course of masonry. One of James Douglass' famous helically framed lanterns is nearing completion.
(Devon Library Service)

Not more than 3 hours at a time could be spent on the rock by the working party. From about three-quarters ebb to about three-quarters flood tide was the limit of their stay, and during that interval the utmost energy of all had to be exerted. With a rough sea, landing on the rock was simply out of the question; but often when at work, the party having perhaps effected an easy landing, the sea would get up, and then it would be necessary for all to seize their tools and hurry off to the boats as quickly as possible. Delay would probably mean being hauled off through the water, for no boat could venture near the rocks with the seas breaking upon them. Occasionally with a smooth sea there is a kind of under swell which breaks with great force on any obstacle interposed in its path. These 'rollers', as they are called, are supposed to be caused by some disturbance in mid-ocean, and at times three or four will follow each other quickly. The look-out man, or 'crow', watches for any indications of the sea getting up or of rollers coming along, and shouts a warning to the men, each wearing a life-belt, have simply to hold on to iron stanchions until they have passed, taking care at the same time that they do not lose their tools.

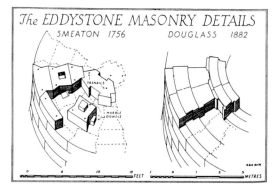

Right: A comparison of the masonry details from the Smeaton and Douglass towers that shows how Smeaton used trenails and dowels, while Douglass preferred dovetailing the blocks of stone together. *(from a drawing by Douglas Hague)*

Work started again on 24th February of the next year. By June the dam was finished and being pumped dry. With the aid of compressed air drills, driven from a motor on the *Hercules*, the gneiss crumbled effortlessly. By August the foundations were ready. The first stone weighing 3¾ tons was laid on 19th August 1879, amid due pomp and ceremony, by the Duke of Edinburgh accompanied by His Royal Highness the Prince of Wales and Trinity House dignitaries. Underneath this block, a bottle containing a parchment scroll detailing the work was set in the cement to record the event for posterity. All the blocks in this lowest course were not only cemented and dovetailed to the reef but also bolted with heavy iron bolts. Four months of toil saw eight complete courses of stone laid by the December when work finished for the year. A total of 131 landings had been made which resulted in 518 hours of work. Following an early start to the next season in March, the solid cylinder of stone was complete by late July 1880 and stood 2½ ft above the high water mark, taking the men out of reach of all but the biggest breakers. Construction proceeded briskly, so quickly in fact that it took just two more working seasons to finish the tower. On 1st June 1881, the Duke of Edinburgh was back on the reef laying the final stone of the cornice, over 130 ft above the waves.

Topped with a lantern house painted red, the pale grey granite tower was a magnificent structure. At 168 ft it was almost twice the height and four times as large as Smeaton's lighthouse, yet it had taken only four years to complete. This was due to several factors; the steam driven *Hercules* which could carry 120 tons of stone at a time, the pneumatic drills for the laborious work of forming the foundations, and the use of mechanical cranes and winches. Using such modern technology, plus quick-drying cement, over ten blocks an hour could be set in place. Following Smeaton' innovation, each one was dovetailed to the next on five of its six faces, requiring an uncommon attention to accuracy from the masons. So good were they at their job it is said that the tower could have been built without the need for mortar. In total, 2,171 blocks of Cornish granite were bound as one into a finger of stone which weighed an incredible 4,668 tons and cost £59,250 – over £18,000 cheaper than Douglass' lowest estimate.

Nine rooms were encased within this giant, the thickness of the walls varying from 8 ft 6 ins at the bottom to 2 ft 3 ins at the top. Working upwards, the rooms were designated as follows; entrance room, engine room, oil room, winch room, battery room, living room, subsidiary light room (to mark an outlier of Eddystone, Hands Deeps, 3½ miles distant), bedroom and service room. Above these came the lantern room originally fitted out with Argand burners and surrounded by 24 6-inch lenses and almost a thousand prisms. This glass structure, magnifying and bending the light rays from the burner, was 10 ft tall and weighed 7 tons, yet revolved without a sound on a bed of stainless steel rollers powered by clockwork. The burners were later to be superseded by paraffin lamps with a mantle, before finally, in 1959, the Eddystone was converted to electricity produced by one of the three generators installed in the tower. The huge revolving lenses were replaced by a smaller optical system, just 3 ft in height, circling a 1,250 watt bulb. To give some idea as to the brilliance of this light, it is calculated at 570,000 candle power (the universal system for measuring lighthouse illuminants and their intensity). One hundred years earlier, the 24 candles of Smeaton's light were estimated at a mere 67 candlepower.

During 1882 the final preparations were being made in the lighthouse. Shelves, lockers, bookcases and bunks were being fitted while surfaces were painted and varnished. A huge open range was installed for the keepers to cook on. Windows were glazed and covered by heavy storm shutters, and the massive gun metal door weighing over a ton was hung. The two fresh water tanks embedded in the solid base, capable of holding a year's capacity of over 4,500 gallons between them, were finally filled. Outside two 2-ton fog bells were suspended from opposite sides of the lantern gallery and connected to the clockwork mechanism that would strike these monsters during poor visibility.

On 18th May 1882, the double white flash of Eddystone's lantern split the night sky over the evil reef for the first time. It was visible for over 17 miles and could reach right into Plymouth Sound. The fifth Eddystone lighthouse was complete and took up the task of its predecessors, a task in which it has never failed and still carries out today. In June 1882, James Nicholas Douglass was knighted by Queen Victoria, "on the occasion of the completion of the new Eddystone lighthouse, with which your name is so honourably connected."

During the winter of 1882 the keepers awoke one morning to find a huge iron cannon, 6 ft long and weighing ½ ton, washed up amongst the rocks at the base of the tower. It was assumed that the heavy seas of the previous night had brought it from deeper water. A closer examination suggested that it had come from the *Winchelsea*, wrecked on those very rocks a few days after Winstanley's tower had been swept away.

Before Douglass and his men left the site for good, the people of Plymouth raised enough money for the Smeaton tower to be dismantled, block by block, and re-erected on the Hoe, standing on a replica of the base left on the reef. This was a fitting end for a structure that had served them so well for so long. It was during this task that William Douglass almost lost his life. The lantern had been removed, and Douglass was standing on top of the truncated stump supervising the numbering and lowering of the blocks. Suddenly, the

Left: This is where Smeaton's tower ended up – in an imposing location on Plymouth Hoe looking out towards the reef on which it used to stand. It is painted in its original colours and is open to the public. *(Author)*

crane they were using failed, and one of the wooden legs caught Douglass and flung him off the edge of the tower, still 70 ft above the waves. Death would have been certain had it not been for the good fortune that brought a wave crashing against the reef, covering the rocks on to which Douglass was plunging, thus breaking his fall. A shocked and dazed Douglass was hauled into a waiting boat but he recovered within a week.

Also during the dismantling William Douglass tells of an incident which occurred during a strong westerly gale. The demolition gang were lodged inside the living room while the storm raged outside. The violence with which the waves struck the base of the tower caused a tumbler of water to be thrown off a table. Such an alarming event only served to emphasise the frailty of Smeaton's light. The tower was dismantled down to its solid stump, which was left with an iron pole planted in its centre, to be colonised by flocks of seabirds which seem curiously attracted to this lonely perch. Despite all the predictions about its disintegration and instability Smeaton's stump still remains today as it was left in 1882. Although discoloured with age, it has not collapsed, and apart from a few blocks in the top courses, still stands to serve as a reminder of the illustrious history of the Eddystone reef. The superstructure of the tower was put together on Plymouth Hoe, painted in bright colours, and opened for public inspection where it can be viewed today, a memorial to the genius of John Smeaton.

The Eddystone continued in its nightly vigil over the approaches to Plymouth Sound without undue incident well into the 20th century. During this time the two giant fog bells were removed and superseded by a modern supertyfon fog horn. However, the most significant changes in the entire history of the Douglass tower took place during the final years of the last century. In common with the other Cornish rock lights, Eddystone has been fitted with a giant helideck on top of the lantern, supported on a lattice-work of steel attached to the top courses of masonry in such a manner that the light from the lantern is unobstructed. Further mention regarding the use of these structures is made in later chapters, but its completion during October 1980 meant that the keepers of Eddystone could now be exchanged every 14 days by helicopter from Plymouth Airport, even in weather and sea conditions which would certainly have caused a postponement of an attempted relief from an open boat being edged between the waiting jaws of Eddystone reef.

Right: When reliefs at Eddystone were done by boat they frequently had to be postponed, even though the relief boat could get to within a few yards of the tower. The keepers in this photo were already 19 days overdue for relief in January 1960. Heavy swell around the reef caused a further postponement and the relief boat returned to Plymouth watched by the three keepers on the set-off. *(Syndication International)*

Even this revolution in operating procedures pales into insignificance with the devastating announcement by Trinity House in March 1981 that they intended to convert Eddystone lighthouse to an automatic station and it would therefore become permanently unmanned. The scheme was to remove the existing optics and navigational aids and replace them with modern silicon chip technology designed for automatic, unattended operations, to be controlled from the Penlee Point Fog Signal Station by telemetry, and subsequently from the Trinity House monitoring centre at Harwich. (Telemetry involves monitoring the functioning of a piece of equipment by means of radio waves.)

In order for the appropriate equipment to be installed, Eddystone's last keepers were withdrawn from the tower on 22nd July 1981. Its warning function was taken over by a lightship named *Eddystone* stationed a mile from the tower which was fitted with a navigational light, fog signal and radio beacon. A quick-flashing lighted buoy with a wave-activated bell was positioned at Hands Deep to take over from the fixed red light in Eddystone's subsidiary light room. By the end of August 1981 the existing lantern and machinery inside the tower had been dismantled and the modern replacements were on their way to Eddystone aboard the THV *Siren* from Southampton. These were transferred from ship to tower by helicopter and lowered through the helideck

Left: The service room inside the lighthouse has this interesting inscription carved into the granite blocks. Psalm CXXVII contains the line; "Except The Lord Build The House, They Labour In Vain That Build It" and the same inscription was also found inside Smeaton's tower. *(ALK archives)*

and lantern roof by a temporary crane. This was an incredibly delicate operation involving the manoeuvring of valuable equipment – some weighing over 500 kg, through the cramped confines of the lantern room and down the twisting stairways to their new sites. One diesel generator was even lowered down the outside of the tower from the helideck and taken inside through the original entrance door.

Once the equipment was actually in the tower, the long job of connecting it all up was started. An incredible 3½ miles of new cables were required inside the tower, in addition to the control switches, fuel and water pipes. The fog signal equipment was delivered by the THV *Winston Churchill* and the period from December 1981 to the end of January 1982 was spent connecting up all the equipment already on station. The new lantern was installed in February 1982 followed by all the radio monitoring equipment, both in the tower and at the shore base 10 miles away. Even though Eddystone would have no permanent keepers it would still be visited regularly by maintenance staff. Accordingly, the tower has been fitted with a modern kitchen, Calor gas cooker, immersion heater, new drinking water tanks, a refrigerator and deep freezer.

Almost all the vital equipment – lantern lights, generators, radio and radar equipment, fog signals, fire detectors, batteries, etc, are duplicated in case of failure. A breakdown or malfunction of any piece of gear is detected instantly and an appropriate signal sent to Harwich. The staff here can also 'interrogate' the systems within the tower to check on their correct functioning. Such is the complexity and sophistication of the stand-by and back-up systems installed in the Eddystone, it is reassuring to learn that a complete failure of the new lantern would be almost impossible. The character of the new light remains the same as the old one – two flashes every ten seconds, but these are now visible from up to 22 miles. Also, the light now flashes continuously, even during the hours of daylight and in good visibility.

It was planned to bring the automatic Eddystone lighthouse back into service on 18th

May 1982 – exactly 100 years to the day since it was first inaugurated. For the week before this date extensive tests were carried out on shore and at sea with the help of the THV *Stella* that assisted in checking the range of the main light, for signal and radio beacons. On 18th May, exactly as planned, the new Eddystone lantern became operational. The Eddystone lightvessel was towed away and the Hand Deeps buoy removed.

Eddystone was Trinity House's first automatic 'rock' station and was accordingly visited on 28th July 1982 by H.R.H. The Prince Philip, Master of Trinity House, whose visit was recorded by a plaque at the top of the stairway. On the same day a circular bronze plaque was erected by Plymouth City Council on the Smeaton Tower on Plymouth Hoe. It bears the following inscription;

EDDYSTONE LIGHT

THIS BRONZE

COMMEMORATES THE CENTENARY

OF THE REKINDLING OF THE

EDDYSTONE LIGHT ON 18-5-1882

ON COMPLETION OF THE TOWER

BUILT BY JAMES DOUGLASS

IT ALSO MARKS THE COMMISSIONING

ON 28-7-1982 OF AN UNMANNED

LIGHT IN THAT TOWER STANDING

14 MILES SEAWARD OF THIS

PLACE ON THE NOTORIOUS EDDYSTONE REEF

IN SALUTEM OMNIUM

While 18th May 1982 was indeed a day for all the Trinity House personnel involved in this mammoth undertaking to celebrate, it was at the same time a sad day in the history of the world's most famous lighthouse. It meant that for the first time since Smeaton's tower became operational in 1759, there would be no keepers stationed on the Eddystone – ending an unbroken run of almost two and a quarter centuries.

Yet this still isn't the end of the Eddystone saga – an era embracing extremes from tallow candles to sophisticated electronic technology. This level of sophistication was further enhanced during 1999 and 2000 when a £350,000 refurbishment programme to convert the lighthouse to run off solar power was conducted. Ninety solar panels were clamped to the outside of the helideck supports in a complete circle below the plane of the light. They generate enough electricity to power the lamp continuously and the fog horn when necessary, and will no doubt soon see off the roar of the diesel engines that previously generated all the energy required.

This is yet another twist in the story of the Eddystone lighthouses, but one that is bound to bring a touch of sadness to all those who knew the Eddystone with its human inhabitants – the many keepers who have done turns of duty within its cramped confines, the fishermen who have exchanged a cheery word with its keepers, the helicopter pilots bringing out the changes of personnel and stores, and not least, the many good people of Plymouth who look for the reassuring double flash far out at sea. They know that the flash they now see is produced from a lifeless stone pillar whose operation is controlled by sophisticated electronics. Although we can be confident that the lighthouse will function in an equally reliable manner, the fact that there is no longer any human life on the Eddystone is one of the sadder aspects that must be endured in the name of progress.

Opposite: The whole of the Eddystone reef and the area around it has become a seething mass of white water in what looks like a south-easterly gale. The helicopter circles the lighthouse before making its approach to land on the helideck. Even in conditions like this when boat reliefs were absolutely out of the question, a helicopter could still change keepers safely. *(Ken Trethewey)*

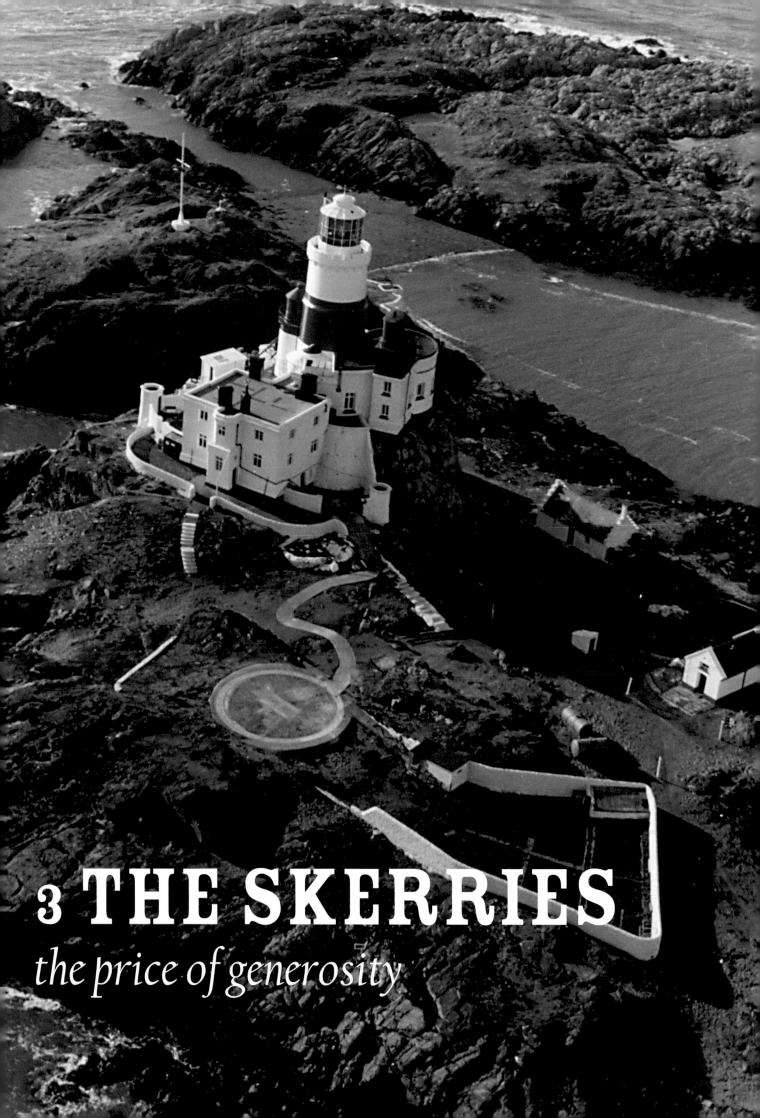

3 THE SKERRIES

the price of generosity

THE DAY Henry Winstanley first put a taper to the tallow candles in his beacon on the Eddystone, a new era of lighthouse construction was about to unfold. He had proved that it was by no means impossible to place a lighthouse on an isolated, sea-swept rock. On the contrary, there was just such a structure for all to see, a tower whose existence large numbers of people were certainly aware of, even if they had not actually set eyes upon it.

Trinity House were doubtless grateful to Winstanley for erecting his structure and for the financial saving it meant to them. They would, however, have been less grateful for the deluge of requests that now followed, in the wake of Winstanley's triumph, for similar beacons to be put up on various equally exposed sites around the coasts of England and Wales. If it could be done on the Eddystone then surely it was possible anywhere. Demands and petitions from various quarters were renewed and presented to Trinity House, who, under the Charter of 1566, had been granted the sole right to supervise the erection of lighthouses and beacons.

In particular, several of these petitions were for the lighting of a notorious group of low-lying, grass-topped rocks 4 miles off the northern coast of Anglesey and 7 miles north-east from the port of Holyhead. They took their name from the Gaelic word *sgeir* meaning reef or rocky islet, and subsequently became known as the Skerries. Their Welsh name is a little more

Left: The Skerries reef and its lighthouse.
(Air Images Ltd.)

descriptive – Ynysoedd y Moelrhoniaid, means 'the Islands of Bald-headed grey seals'. What is interesting about the lighthouse on this site is not so much the tale of its construction, but more the history of how it came into existence and also the curious events which took place before it was finally acquired by Trinity House from private ownership.

As long ago as 1658 there were rumblings of discontent about the Skerries from merchants trading between Great Britain and Ireland. Chief amongst the protesters was one Henry Hascard, a private speculator, who highlighted the need for some kind of beacon on these rocks, particularly as they were in the direct path of vessels plying between Liverpool and Dublin. He appealed to Oliver Cromwell's Council of State and further offered personally to erect a beacon there. Trinity House, jealously guarding their rights, opposed Hascard vehemently and the matter lapsed, even though in April 1662 they agreed in principle that the construction of a beacon on the Skerries was desirable.

A particularly high-profile casualty of the Skerries at this time was the first British 'Royal Yacht' *Maria* – an eight gun ship that was a gift to Charles II in 1660 from the people of Holland. On 25th May 1675 she was en-route from Dublin to Chester with 46 passengers and 28 crew when she struck the Skerries in full sail and dense fog at 9.30pm. She came to rest on her side against one of the smaller rocks with her mast touching land, enabling just over half the passengers to escape to safety along the mast-bridge. The ship's Master, his bosun and another thirty or so souls perished.

With the dawn of the 18th century came renewed attempts to get this cluster of rocks lit. Over 140 merchantmen signed a petition to this end in 1705. It had been drawn up by Captain

Opposite: The Skerries, as seen from an approaching Trinity House helicopter. The helipad was built on part of a walled garden in which the keepers grew vegetables. The part that survives is still under cultivation in this shot.
(C.J. Foulds)

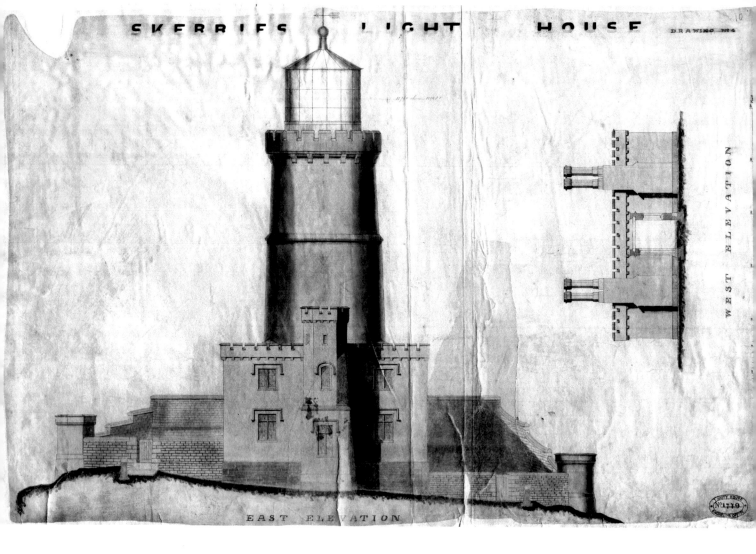

EAST ELEVATION

WEST ELEVATION

Above, and opposite: Two of the original hand-painted Trinity House plans for their new lighthouse on the Skerries.
(Trinity House)

John Davison who duly presented his signatures to the Attorney General in 1709, pointing out that it was only because, "…many ships were cast away…chiefly for want of a light in the night on the Welsh Coast" that his petition was necessary. The Attorney General at that time was Sir Edward Northey who, out of courtesy in 1711, invited the views of the Elder Brethren of Trinity House on this matter. Again they trotted out the argument that Queen Elizabeth I had granted to them the sole right in these matters, and under no circumstances were they prepared to put the undertaking into the hands of private individuals. Also, they now saw no pressing need for a light on this site. However, if those involved were still concerned about the Skerries and were prepared to meet some of the cost, then Trinity House would build a lighthouse there.

This was indeed a curious statement to make, particularly as in 1662 they were in agreement with the principle but the construction, they said, was out of the question owing to the isolated and exposed position of the rocks. Yet now, just 49 years later, the Elder Brethren were actually offering to do the building themselves. Could it be that the successful placing of two lighthouses on the notorious Eddystone had altered the perspective with which they viewed isolated sea rocks, and that sites previously considered impossible were now being regarded as a more feasible proposition? If this is so then Rudyerd and Winstanley had certainly achieved far more than simply illuminating the Eddystone.

What is probably the most significant decision in the whole history of the Skerries and their lighthouse was now taken by Northey and his Law Officers. They disagreed with Trinity House about their sole rights, and recommended to the Crown that Captain Davison's petition and offer of construction be granted in 1710. Legal difficulties encountered when trying to negotiate a lease of the Skerries from their owner, one John Robinson, meant that

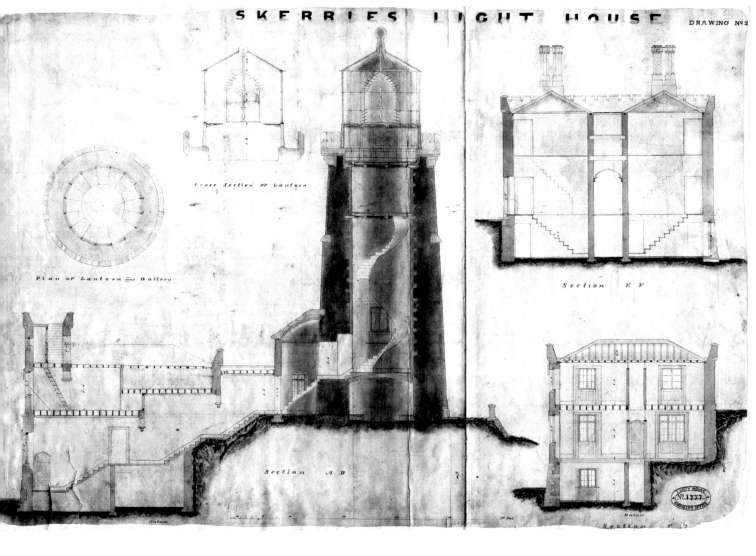

Davison and his fellow petitioners never availed themselves of this historic decision and their interest waned in 1712.

Shortly afterwards, in June 1713, the Skerries had been acquired from John Robinson by a wealthy merchant, William Trench, with a 99-year lease on payment of £10 rent for the first year and £20 for subsequent years. By 13th July of the following year he had added to this a patent for the erection of a lighthouse, financed from his own pocket, at an additional annual rent of £5, but with a provision to collect dues of 1d and 2d a ton from shipping for the 60 years after its completion.

Trench could not have had a more disastrous start. In 1714 he supervised the loading of the first boat with men and materials and watched its departure for the Skerries. In this boat was his son and six workmen. William Trench was never to see his son again; before the party could reach the rocks their boat was wrecked in a freak storm which swamped the heavily loaded boat and snapped its mast. It was driven on to Platters Rock with the loss of all seven men. As might be imagined, such events were a hammer blow to his enthusiasm and it was not until 1717 that he finished his lighthouse. It had cost him the life of his son and in excess of £3,000. He had produced a tower, "about 150 foot higher than ye sea about it and on ye 4th November a fire was kindled therein and ever since supported." It was a landmark in more ways than one, for apart from the warning it gave of the Skerries it was also the first permanent light along the entire west coast of England.

Rather than relying on candles for illumination Trench had installed a coal-burning grate in the lantern. His reasoning was not quite as absurd as it might first appear. The northern coast of Anglesey was frequently engulfed by notorious sea fogs which can form in a matter of minutes and last for several days. These swirling mists were the cause of many fine vessels prematurely ending their days on these rocky outcrops. It is unlikely that any source of illumination could pierce far into these clinging fogs, but of the choices of illuminant Trench could reasonably have made, a cast-iron grate 3 ft across piled high with burning coals was perhaps the most satisfactory compromise.

Left: The gulls are circling the Skerries lighthouse as a heavy squall, complete with rainbow, moves past in the background. On the left of the picture is one of the stepped gable ends of the original keepers cottage.
(C.J. Foulds)

Its intensity, although completely inadequate by later standards, would certainly have been an improvement on a handful of tallow candles.

According to contemporary documents, the circular stone tower was about 36ft high with the grate set at 78ft above the high water mark. At a later date, but on the same islet, a lightkeeper's cottage was built from the local stone with the characteristic Anglesey feature of stepped gable ends. This structure still exists today and has recently been renovated. Although it had ceased to be inhabited since the middle of the 19th century, it is probably the earliest remaining purpose-built offshore lighthouse keeper's accommodation structure anywhere in the world.

Large amounts of fuel were required to keep the light in service, 80 tons for an average year, and upwards of 100 tons during years with severe winters. It was stockpiled on Carmel Head and brought across to the Skerries by boat. The smoke from the smouldering coals soon became a serious problem, particularly when the air was still with no breeze to remove these

Right: An dramatic shot of the Skerries at night, showing not only the intensity of the main beam, but also the fixed red sector light to the left of the main tower.
(C.J. Foulds)

vapours from obscuring the glow. So frequent was this occurrence that it was not long before the Skerries' coal grate earned for itself the reputation of being one of the worst lights in the United Kingdom.

On 21st June 1725, 8 years after its completion, William Trench died. The ownership of the Skerries was passed to his wife Ruth, while the lease of the lighthouse passed to his daughter Anne and her husband Sutton Morgan who was subsequently to sell it for a nominal sum. The reason for this unhappy occurrence lay in the fact that the Trench family experienced difficulty with the enforcement of light dues, particularly in the port of Liverpool where the majority of traffic passing the Skerries was bound to or from. Losses were estimated at over £100 a year during the infancy of the light, which coupled with the £3,000 spent on construction, meant that for most of the 12 years preceding his death Trench was an impoverished and broken man.

Whether it was out of sympathy for his descendants or for some other reason, Trench's family, upon presentation of the accounts for the lighthouse, were fortunate in being granted the lease of the light, together with the right to keep all the dues, *in perpetuity*, by an Act of Parliament of 24th June 1730. This was an exceptional act of generosity and a precedent they were later to regret, one that was to give the Skerries a unique place in lighthouse history.

It was during the coal-burning years on the Skerries that an amusing incident is related about this light. In the late 1730s the brazier was attended by a married couple who received £15 annually for their labours. One night in 1739 the keeper and his wife were busying themselves with their duties when they were interrupted by a knock at the door of their cottage. Knowing themselves to be the only inhabitants of the island they put their experience down to a trick of the wind. Upon hearing a second knock they were somewhat perturbed. The worried discussions that followed were interrupted by the door swinging open to reveal a naked negro. The wife was immediately seized by hysteria, swearing they had been visited by the devil, whereas when the truth became known, they were actually face to face with a luckless survivor of a recent shipwreck on the rocks, about which the keeper was oblivious owing to the heavy mists.

In 1759 Trench's original tower was demolished by Sutton Morgan's son and replaced by a new stone tower 28ft high and 22ft in diameter, together with a new jetty for landing coal. It had a new iron grate which increased its range to 15 miles. In 1778 the Skerries lighthouse became the property of Morgan Jones Snr, the High Sheriff of Cardiganshire who managed his property with particular diligence. By 1803, however, the Skerries coal fired grate was beginning to look distinctly dated compared to recent innovations in lighthouse construction and illumination. Trinity House 'advised' Morgan Jones – because that was all they could do – that extensive repairs and upgrades were required. Wisely, he listened to their advice.

The Skerries patent was amended to substitute all references

Left: Repainting a lighthouse on an exposed site isn't just a case of adding an extra top coat! This rare shot shows the Skerries being repainted prior to automation which involves removing every layer of paint down to the stonework, and then repainting from scratch.
(C.J. Foulds)

to a coal fired grate with oil lamps. Samuel Wyatt was the Trinity House consulting engineer and he produced plans to raise the tower by a further 22 ft, complete with battlements along the top. Above this a proper iron-framed and glazed lantern, 12ft high, was installed, complete with 16 Argand lamps and reflectors. They gave a range of approximately 18 miles for the light and were lit for the first time on 20th February 1804.

It was also in the 19th century, particularly the early half, that trade between England and the Americas expanded considerably. So much so that, in 1834, it was calculated that the profit alone for the Skerries lighthouse, after expenses and maintenance costs had been deducted, was well over £12,500. This was due in no small part to much of this trade being conducted through the thriving port of Liverpool, causing all such vessels to pass the Skerries and therefore render themselves liable to light dues. In less than a century the coal-burning, smoke-enshrouded light which had bankrupted its builder was transformed into a highly profitable, oil-burning lantern, a fact that had not gone unnoticed by Parliament.

Exactly how profitable they were not entirely sure to begin with. By 1834 the light was in the hands of another Morgan Jones who had inherited it from his father of the same name. He was reluctant to produce accounts for the light, as demanded by a Parliamentary Committee of Enquiry in that year, claiming immunity from so doing by an earlier Act passed in the days of George II. When the figures were eventually extracted from Jones, the government were staggered to find that in addition to his annual profit of over £12,000, the owner was also receiving £1,700 from them under a reciprocal agreement made during the earlier history of the light when it was incapable of recovering its own costs. Such an absurd state of affairs would not exist, Parliament said, if all lighthouses around our coasts were managed and administered by one responsible body, whose job it would be to levy a standard rate of light dues for all lighthouses. This would do away with private lighthouses whose owners were at liberty to charge whatever they thought appropriate in the circumstances, which had led to unscrupulous profiteering by the majority of these fortunate owners.

As a result of this proposal the Act of 1836 was passed, *An Act for vesting lighthouses, lights and seamarks in England in the Corporation of Trinity House*, which gave Trinity House the

Right: Principal Keeper W.J.Hast going ashore from the Skerries lighthouse on board a Trinity House vessel. (*Caxton Press, courtesy of ALK archives*)

authority to purchase any remaining private lights – by compulsory purchase if necessary – and bring them under their jurisdiction. This legislation was the start of lengthy legal wranglings over the Skerries which were to further guarantee for this beacon a special place in the annals of lighthouse history.

Trinity House was naturally keen to acquire the ownership of this lucrative source of income as soon as possible, yet Morgan Jones was a resolute man who was equally determined not to part with his property without a fight, particularly as it was now earning him over £20,000 a year. He rejected offers of first £260,000, then £350,000, and lastly £399,500 for his lease. For five years Jones resisted the pressure of the Elder Brethren, taking the line that his family had been granted the right to all the light dues from the Skerries "in perpetuity" and therefore the Act of 1836 did not apply to him. He would doubtless have done so for a good deal longer had it not been for his untimely death in 1841.

Here was the ideal opportunity for Trinity House to take control of the light they had fought over for so long, yet the battle was not quite over. The executors of Morgan Jones, the same solicitors who had been fighting for him against Trinity House, insisted on the final settlement being decided by a jury. This was a shrewd legal move which was to reap the intended rewards. It led to a jury sitting before the High Sheriff at Beaumaris, Anglesey, on 26th July 1841, a jury who awarded to Morgan Jones' estate the phenomenal sum of £444,984.

The fate of the Skerries lighthouse was finally sealed. From that day forward the light was maintained by Trinity House who for the privilege had to pay a vast sum, sizeable even by today's inflated standards, yet undreamt of in 1841. The Skerries was the final private lighthouse

Left: After automation
the light flashes
continuously, 24 hours
a day.
(C.J. Foulds)

to pass into public control and fetched a King's ransom for doing so. The beacon which had previously gained fame in 1777 for being one of the worst lights in the United Kingdom and leaving its builder a destitute bankrupt had now sprung to prominence for the second time in its history. How Parliament must have bitterly regretted their generous decision in favour of the Trench family over a century earlier.

Within three years of their purchase Trinity House and James Walker, their consulting engineer, had started planning further improvements and modifications. In the middle of 1845 a start was made on another new tower 75ft high with a ring of even more impressive castellations around the top. A new iron lantern, 14ft in diameter and 16ft high, contained 16 Argand lamps with mirrored reflectors that were revolved by clockwork. Its light was set at 119ft above high tides, had a range of 18 miles, and was first seen on 23rd September 1846.

Apart from a continuous process of modifications and improvements in line with the technology of the day, this is the structure that is found on the Skerries today. A huge first order Fresnel lens from Chance Brothers that revolved around a new Fresnel oil lamp on a clockwork pedestal arrived in 1851, increasing its range to 21 miles. A fog horn was added in September 1876, and a new semi-circular lantern 10ft in diameter and 21ft high was attached to the south west of the existing tower in 1903. The light it contained shone a fixed red beam, visible for 14 miles over a cluster of dangerous outliers to the Skerries, although this light was discontinued in the 1980s. Electricity arrived in 1927 and automation in 1987 ending a period of 270 years of continuous manned service at the same lighthouse.

These 270 years have not been without incident, indeed 'troubled' might be a fair description of the history of the Skerries lighthouses. However, throughout the embittered legal complexities and the stubborn resistance of all the various parties concerned, one fact mustn't be overlooked. The disputes surrounding its acquisition from private hands were over a lighthouse – tangible evidence of one of mankind's more humane instincts, and a building whose sole purpose was to save lives. As such its purchase would have been justified at twice the price, for no value can be placed on the lives which have been spared because of its existence.

4 THE SMALLS
a lighthouse on stilts

THE WILD coastline of Wales has claimed many fine ships which failed to accurately negotiate the innumerable hazards that bedevil the water around this historic kingdom. It is hardly surprising, therefore, that for the subject of this chapter we remain in Celtic waters. In the south western corner of Wales, the old county of Pembrokeshire offers breathtaking scenery. It is an area of contrast; towering sandstone and limestone cliffs rise from azure waters, while jagged promontories enclose wide sandy beaches. Seabirds, notably thousands upon thousands of fulmars, guillemots and razorbills, are drawn to this temperate corner, where they nest undisturbed on the narrowest cliff ledges or offshore islands, both blanketed with a lush carpet of colour during the spring and summer months. Much of this glorious coastline is part of the Pembrokeshire Coast National Park, ensuring the conservation of its attractions for future generations, despite the recent encroachment into Milford Haven by

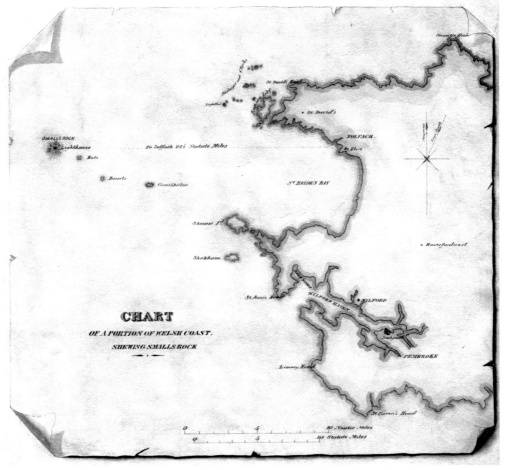

Left: A hand-coloured Trinity House plan showing the location of the Smalls lighthouse in relation to the Pembrokeshire coast. *(Trinity House)*

rows of oil storage tanks, jetties and a handful of oil refineries which despoil this huge inlet, once described as, "the finest natural harbour in the world."

This rugged coastal scenery marks the westernmost extremity of Wales. Travel further west and the next landfall is the southern coast of Ireland. As the fertile soils of Pembrokeshire plunge into the sea it is with some reluctance that this wild landscape takes its final bow. Stretching seaward from St Ann's Head for over 20 miles is a string of islands, each a barren, windswept refuge for multitudes of seabirds, the final remnant of a Celtic landscape struggling to hold its own above the waters of St George's Channel. First comes Skokholm, then Skomer, both much beloved by ornithologists, beyond which rises the rounded profile of tiny Grassholm with its thriving gannetry. Next in the chain are two lonely clumps, 2½ miles apart, that barely struggle above low water, known as the Hats and Barrels reefs. Finally, 21

Opposite: The Smalls reef is an extensive jumble of rocks spread over several miles at the entrance to St George's Channel. Most of reef lies just below the waves and its lighthouse sits on one of the two biggest rocks. *(Author)*

miles from the nearest land, is a tiny cluster of rocks known collectively as the Smalls, the setting for this chapter and the site of a lighthouse that boasts the dubious distinction of being the most remote station under the jurisdiction of Trinity House.

Imagine, if you can, the sheer isolation of this lighthouse – over 20 miles from the nearest Pembrokeshire cliffs and even further from any sizeable settlement. Completely uninhabited, these 20 jagged rocks are assailed from all points of the compass with unequalled ferocity by wind and wave. They are over twice the distance the Eddystone is from Rame Head, and break the surface over a distance of 2 miles at the entrance to St George's Channel, straddling the path of the busy seaway through the Irish Sea to Liverpool. There can be few more God-forsaken sites around our entire coastline. The fact that these rocks lie in one of the busiest sea routes in the North Atlantic only serves to underline the malevolence of the reef. Rising to 12 ft above high spring tides, they provided a reasonable landmark in clear, calm weather, yet when Atlantic swells over-topped these obdurate clumps the foaming swirl of broken water was the only clue to the whereabouts of this hidden menace.

There have been two lighthouses placed upon the Smalls; the original was the setting for one of our classic lighthouse sagas, events which have been immortalised by various sources and have passed into the folklore of this corner of Pembrokeshire. I make no apologies for retelling it here, but first, some background to the origins of this historic structure would be in order. Again we must look to the 18th century when the Smalls were owned by John Phillips, a Welshman and native of Cardiganshire, who was appointed manager of Liverpool's North Dock in 1770. Why anyone would want to 'own' a series of wave-lashed lumps of rock in the middle of the Irish Sea has never been fully explained. Whatever the reasons, Phillips owned the Smalls reef. He had watched the construction of lighthouses and beacons around the expanding port with interest, and was naturally keen to mark his own personal reef in some way similar. Unfortunately being plagued with near bankruptcy he was not particularly well placed to push his creditors for finance. The events of a winter's night in 1773 spurred him into action.

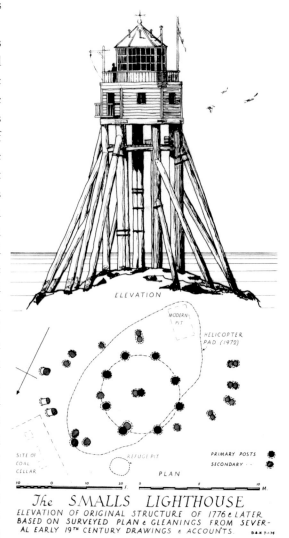

ELEVATION

The SMALLS LIGHTHOUSE
ELEVATION OF ORIGINAL STRUCTURE OF 1776 & LATER
BASED ON SURVEYED PLAN & GLEANINGS FROM SEVER-
AL EARLY 19TH CENTURY DRAWINGS & ACCOUNTS.

Left: An extremely accurate drawing by Douglas Hague of what Whiteside's lighthouse would have looked like after many extra oak struts had been added to strengthen the original nine.
(D. B. Hague)

During a wild night late in that year, the clipper *Pennsylvania* of 1,000 tons was caught in the grip of a tearing gale as she voyaged from Philadelphia to Bristol. After a peaceful Atlantic crossing the seas had suddenly been whipped into a frenzy by a storm which tore at the very heart of the vessel as she pitched helplessly in the heaving swell. Only a handful of miles from her destination, the mountainous waters dwarfed the tall-masted ship which was picked up like driftwood and hurled on to the largest rock of the Smalls. For a week afterwards the stark reminders of the tragedy were washed ashore in tranquil Pembrokeshire bays; timbers, rigging, shredded sails, and the corpses of a few of the 75 souls who had perished on the reef. Enough was enough. Phillips, reputedly a passenger on the ill-fated vessel, was not insensitive to the public clamour and vowed to mark in some way his own personal maritime graveyard.

Left: 'The Intrepid and The Smalls' is a vivid watercolour by the artist Simon Swinfield of Solva of what Whiteside's lighthouse might have looked like in the days of sail.
(Simon Swinfield)

With an air of bravado he boasted of his intention to build a lighthouse, "…of so singular a construction as to be known from all others in the world as well as by night as by day." Further, more precise, details of his "great and holy good to serve and save humanity" were still somewhat vague, yet this advance publicity obviously had some effect. On 4th August 1774 the Treasury granted him a lease to erect a beacon on the Smalls rock. Plans were now invited and closely scrutinised by Phillips who eventually settled on the design from a 26 year old man called Henry Whiteside. Following the contemporary trend in lighthouse construction, Whiteside had no previous experience in the field but was engaged in the unlikely profession of "a maker of violins, spinettes and upright harpsichords", although he had undergone training in wood carving. Exactly how these almost absurd qualifications got him the job isn't certain, yet there is no dispute over the fact that Whiteside planned and built a beacon of wood which subsequently stood for well over three-quarters of a century on that miserable outpost of civilisation.

His design was unique. There had never existed a lighthouse of such a curious form before, and there was never another built in the mould of the original. In essence, Whiteside raised an octagonal wooden structure, crowned by a lantern, on sets of oaken stilts. The house was 15 ft in diameter and divided into two sections, one above the other. The lower served as sleeping and living accommodation while over this, and connected to it by a trap door, rested the lantern room, surrounded on the outside by a gallery. Beneath the living quarters there was also a kind of landing, open to the elements, on to which casks and boxes of non-perishable goods could be lashed. Windows were let into several of the sleeping quarter walls, so too was a door which led to the outside balcony for "…keeping bad weather watch, and also to keep the windows of the lantern clean and polished bright." Several chimneys projected through the roof of the lantern house to conduct away candle smoke, and the balcony was fitted with the obligatory flagpole and crane hoist. A rather ungracious account once described the structure as a "strange wooden-legged Malay-looking barracoon of a building." (A barracoon was a stilted enclosure used for storing slaves on the west coast of Africa prior to their sale.)

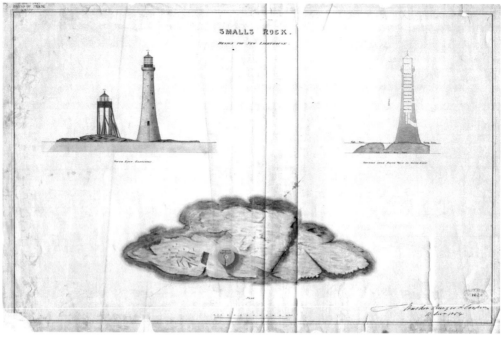

Right: On a plan dated December 1854 called 'Design for New Lighthouse' we can clearly see a comparison of Whiteside's wooden light and the masonry tower that James Walker designed and James Douglass built. *(courtesy of Trinity House)*

Although of unusual appearance it was not a design of pure fancy. Whiteside argued that instead of resisting the force of the waves, his plans would allow the greatest weight of water to pass through the supporting piles unhindered, as these would offer little resistance. His logical and practical design clearly impressed Phillips. By June 1775 he had amassed sufficient funds for materials to be purchased and a start made. A later newspaper report of 1899 made the following observation: "His undertaking was a sudden transition from the sweet and harmonious sounds of his own instruments, to the rough surging of the Atlantic wave, and the discordant howling of the maddened hurricane; and from the fastening of a delicately formed fiddle, to the fixing of giant oaken-pillars in a rock as hard as adamant!"

Although the Smalls were such a vast distance from land, Whiteside did not adopt the procedure of lodging in a nearby floating barrack which was later to become standard practice at isolated lighthouse sites. Instead, he established a base at Solva (sometimes referred to as Solfach on early maps), a tiny fishing village nestling in a sheltered creek over 22 miles from the Smalls. It would be difficult enough to build anything on these venomous humps, but Whiteside certainly was not making his task any easier by necessitating such a lengthy row before a landing could even be attempted. Often it was impossible and the work party had to return without lifting a tool.

Soon after mid-June 1775 he left for the rock in a cutter, accompanied by a gang of rugged Cornishmen, eight miners, one blacksmith and two labourers. They lost much time during this first visit when "…the sea being turbulent, and a gale of wind coming on suddenly…" caused the boat to lose its moorings. They did have time, however, to choose the largest islet of the group "…always above the water, about the bigness of a long-boat," as the site for the new lighthouse.

Things went badly for Whiteside and his artificers during the early days. On their very first landing, while the men set about excavating the holes for the wooden piles, the winds freshened and the sea began to break over the rock. The cutter rode uneasily at anchor half a mile away and soon had to cast off to run for the open sea when the storm freshened further, giving no time for the work party to be brought aboard. The gale turned the sea into a boiling fury, preventing the boat's return for two whole days. When conditions had moderated sufficiently for an approach to be made, the crew found the gang of luckless, yet fortunate men gripping a solitary iron rod they had managed to drive deep into the heart of the reef before the storm. For two days they had survived, without food or drink, clinging to a solitary post with numbed hands while huge seas broke over them. This was a forceful reminder of the perilous nature of their task, yet Whiteside and his gallant Cornishmen were made of stern stuff and stuck at their work doggedly.

By early October the weather was deteriorating, forcing a halt to the proceedings. In the four and a half months they had been active, only nine whole days had been spent on the rock. During this time a hole 18 ins deep had been sunk for the central pillar, a second hole started, the positions of the remaining piles marked, and a hut to house 12 men for a fortnight put up. It had not been easy; from the outset the elements had resisted many of their attempts to land, the long and frequently abortive journeys from Solva were both exhausting and dispiriting. Working conditions were positively evil, at the mercy of wind and wave, struggling to keep upright, while at the same time swinging heavy implements. It was reported that:

Left: Probably a 1960s' view of the Smalls lighthouse showing the interesting lantern roof fixtures and fittings – all of which were removed when its helideck was fitted a decade later. *(Ken Davies)*

...the clothes of the workmen were often washed off in shreds from their bodies, and their skins were severely lacerated by falling heavily on the pointed surface of the rock. The workmen, on such occasions, could only save their lives by lashing themselves to the iron bars and ring bolts secured in the rock for that purpose.

During the winter Whiteside erected the whole structure in a field at the top of Solva harbour. This was a particularly fortunate piece of foresight as the original design called for the building to be supported on nine cast-iron stanchions of sectional construction. Upon completion, these proved to be weak at the joints and therefore unsatisfactory. Whiteside amended his plans and substituted oak for metal, a change which necessitated an extensive search through the forests of Wales for specimens of suitable length and straightness. The wooden legs raised the living quarters and lantern to 65 ft above the level of the rock. They were arranged in a circle around the central pillar spanning 40 ft, with individual diameters of about 2 ft.

Work was resumed in the spring of 1776 and, thanks to his diligence during the winter, little trouble was experienced in consolidating the structure on the Smalls. In fact, so speedy was the construction, it was complete by the end of August and lit on the first day of September. It became immediately distinctive because the lantern, containing four lamps with glass reflectors, showed two colours – the main fixed white light, visible for 12 miles, was topped by a less powerful green beam which was only visible at short range and from certain directions. This, no doubt, was Whiteside's attempt to mark the passage between the two clumps of rocks, although any vessel taking this course was surely tempting fate. Before the jubilant gang of Cornishmen deserted the rock for good, a cellar and store, 10 ft by 6 ft by 6 ft, was hewn out of the rock for the storage of certain weighty commodities such as the drinking water or 12 tons of coal consumed annually which would be impractical to store in the beacon.

Left: Principal Keeper Terry Cresswell and Assitant Keeper Neil Hargreaves examine what remains of the huge oak stumps from Whiteside's wooden lighthouse in 1977. Most of the stumps were covered with the slab of concrete seen here when a helipad was constructed on the reef before the tower had a helideck.
(C. J. Foulds)

The Smalls were at last lit by a lighthouse which could quite rightly justify Phillips' claim to be "of so singular a construction as to be known from all others in the world." Its stilted form and bi-coloured lantern made it unique. In rough weather the Atlantic broke against the rock and rolled through the legs of the structure before returning to a foaming sea. The waves were ruthless in their treatment of the beacon. Such an exposed site is regularly subjected to mountainous seas crashing against the unyielding rock. By December 1776, under four months after it came into service, the situation on the Smalls was far from good. The keepers told how the light swayed drunkenly when hit by large waves, so Whiteside was summoned in order that strengthening could be carried out. Without it the life of his beacon was in serious danger.

In January 1777 he, together with his blacksmith, left for the Smalls to effect repairs. These would be further oak piles leant against, and joined to, the existing nine, thus increasing the stability and making the legs more able to resist the lateral force of the waves. While the two men were on the rock one of the almost continuous winter gales blew up, stirring the sea into a frenzy around the solitary wooden refuge. For over two weeks the storm raged, making it impossible for Whiteside to leave the lighthouse. There were normally only two keepers on duty, but with the two extra inhabitants supplies quickly dwindled. It was not long before the situation became desperate. The only method of communication with the mainland was by boat. The weather prevented use of this so Whiteside took action that has now come to feature heavily in tales of desert island 'castaways'. He wrote a message, sealed it in a bottle, put this in a wooden cask, and flung it into the sea, leaving fate to do the rest. In fact, he made three identical copies, one of which arrived on Newgale Sands on the Pembrokeshire coast, the second beached itself on the shores of Galway Bay, Ireland, and the last floated in the opposite direction altogether to come ashore near St David's. On the outside of the cask were the words, 'Open this and you will find a letter'. Inside was a note addressed to Thomas Williams, John Phillips' local agent for the light, the contents of which explained the desperate plight of the marooned men:

To Mr Williams,
Smalls, Feb 1st, 1777
Sir. – Being now in a most dangerous and distressed condition upon the Smalls, do hereby trust Providence will bring to your hand this, which prayeth for your immediate assistance to fetch us off the Smalls before the next

spring, or we feel we shall perish; our water near all gone, our fire quite gone, and our house in a most melancholy manner. I doubt not but you fetch us off from here as fast as possible; we can be got off at some part of the tide almost any weather. I need not say more, but remain your distressed Humble servant,
HY. WHITESIDE.

A further plea from the two keepers and blacksmith was attached at the bottom:

We were distressed in a gale of wind upon the 13th January, since which have not been able to keep any light; but we could not have kept any light above 16 nights longer for want of oil and candles, which make us murmur and think we are forgotten.
ED. EDWARDS.
GEO. ADAMS.
JNO. PRICE.
We doubt not but that whoever takes up this will be so merciful as to cause it to be sent to Thos Williams, Esq, Trelethin, near St David's, Wales.

By sheer good fortune it was found by fishermen at almost the exact location it was to be delivered to. At the earliest opportunity a boat was despatched to bring the weary and hungry captives off the Smalls. They all survived and were none the worse for their ordeal, yet this period of constantly appalling weather had taken its toll on the lighthouse.

Further repairs were vital for the safety of the beacon, but John Phillips was again short of capital. The keepers were dismissed and the Smalls once again plunged into darkness while a group of Liverpool businessmen took over Phillips' interests. As most of their trade came into the city on boats, and a large majority of these passed the Smalls, they wisely saw the need to return the light to service. Influenced by the work Phillips had done in establishing the station in the first place, on 3rd June 1778 he was granted a 99-year lease from the traders at an annual rent of £5, who also advanced monies for the addition of further strengthening struts. By September of the same year a new white light radiated from the Smalls, produced by four 5 ft 6 ins reflectors, once again warning ships of these perilous rocks.

Right: The author descends the dog steps from the entrance door of the lighthouse onto the reef. The massive stepped base of the tower is really impressive from this position.
(Author)

Sometime during the period 1800–1801 an incident took place on the Smalls which had far-reaching effects on the manning of rock stations in the future. There were, as normal, two keepers on duty in the light. Conditions outside were particularly atrocious that winter; continual storms assailed the rock extending the normal duty of one month into a lengthy exile of four months for the two men. The pair of unfortunates were Thomas Griffiths and Thomas Howell, both well known characters in Solva, their home. What these two loved more than anything else was to argue. Every second they were in each other's company was spent bickering about topics so wide and varied, it was said there was nothing they had not disagreed about. They could empty bars of public houses with the force of their arguments, especially when it looked as though they were coming to blows over something. They never did. However much they raged, the towering Griffiths and middle-aged Howell had never been seen in physical conflict. It was perhaps just as well, for

1861
(Courtesy: Trinity House)

c 1977
(C. J. Foulds)

1984
(Author)

2005
(Steven Winter)

Above: An interesting comparison of how the same lighthouse on the Smalls has changed over the years from a plain masonry tower, that was subsequently painted with red and white stripes, and then had a white helideck added, before the stripes were removed and the helideck was painted red.

Griffiths would not have had the slightest difficulty in despatching the weaker man. They were now marooned together on the Smalls where they could squabble to their hearts' content and nobody would have to listen.

During their prolonged duty, Griffiths suddenly and unexpectedly died, collapsing in the lantern room and striking his head on a metal stanchion as he fell. Thomas Howell, not wishing to be accused of murdering his companion, did not commit the body to the sea but made instead a shroud for the corpse and installed this in a crude coffin fashioned from a cupboard. This box was then lashed to the rails of the outer gallery in an attempt to lessen the stench from the decaying body. Almost out of his mind with fear, Howell managed to raise a flag of distress and make appropriate entries in the log about his companion's death. All he could do now was to wait for his long-overdue relief and pray to God that it would not be long in coming.

The storm refused to abate, if anything it got worse, driving the waves high into the air to crash down on the lantern gallery, sending sickening tremors through the creaking house. One monstrous breaker split Griffiths' coffin in two and tumbled the enshrouded corpse on to the gallery where it rested with one arm caught in the railings. Many times boats set out to relieve the keepers, only to be beaten back by boiling seas. On one occasion a vessel managed to battle near enough to the rocks to notice a man leaning motionless on the gallery next to a flag of distress, his hand waving freely as if trying to attract attention. Upon reporting their findings, some concern was felt for the keepers, yet every night, as punctual as ever, the glow of the lighthouse could be seen from the land. What then, could be the matter on the Smalls?

During a respite in the storm a boat raced out from Milford to snatch off the two men. Only then was the terrible truth realised. In a state of complete exhaustion, near hysteria, and absolute fatigue the solitary keeper had never failed in his duty, tending the lantern for day after miserable day with only the putrefying remains of his colleague for company. His ordeal had affected him so greatly it was said that some of his friends did not recognise him on his return. As the events became known, it was obvious that what the earlier rescue bid had seen was the free hand of the corpse being blown by the wind in a morbid greeting, enticing them to come nearer so that they too might join him in eternity. After the knowledge of this story became widespread it was made standard practice for three keepers to be assigned to isolated lighthouse stations, so that if one keeper should fall ill the light could be maintained by the other two without undue suffering and so prevent a repetition of the Smalls' tragedy.

Although it performed an admirable function, there was no doubt that Whiteside's structure suffered much damage during times of storm. The total height of the whole tower was only 72 ft, which made it by no means unassailable for large rollers. The force of water constantly surging around and through the piles made the lantern and living quarters rock giddily, sometimes as much as 9 ins out of perpendicular. Each winter brought a fresh crop of problems and repairs to ensure the continued service of the lighthouse.

A storm of exceptional severity in October 1812 closed the tower until the following spring. Whiteside describes the events of 18th and 19th October in a letter to Robert Stevenson:

…It was a tremendous storm here such as cannot be remembered. The Smalls has now been up 37 years with no damage: only a few panes of glass broke sometimes [sic]. The men suffered very little hardships, only being frightened at the time of the storm. If they had stay'd there long their house would have been somewhat leaky the windows being broken. They had plenty of firing and everything they wish'd for to live upon. One of the men lived there 13 years and they are going there again as soon as it is made tenantable.

Not only did it break windows but the lantern was blown completely off. Remnants of this and the actual planking of the house were washed ashore, and only when this debris was delivered to Whiteside's door in Solva was he aware that anything was amiss on the Smalls. In fact, the situation was rather more serious than the "somewhat leaky" condition described by Whiteside. The whole building was obviously breaking up. The oak legs were snapping like matchsticks, tearing huge chunks of the floor from the lower room with them as the sea wrenched them from the rock. The breakwater built around the base of these to prevent such a thing happening had

been destroyed, probably by a single wave, and the interior of the living quarters was running with water which was cascading in through the gaping holes in the roof and walls. The terrified keepers despatched urgent pleas for assistance in casks, exactly as their forerunners had done earlier, but it was not until 8th November that a boat, the *Unity* containing Whiteside and Captain Richard Williams, could leave for the Smalls. They found that the keepers, huddled together in the rickety remains of the lighthouse, had existed throughout their three weeks' ordeal on bread and cheese alone.

Left: An unusual view of the Smalls lighthouse. By lying on the helideck and pointing his camera downwards during a storm a former keeper has caught the moment when a wave sweeps over the reef and swirls around the base of the tower. The landing stage where the a relief launch could tie up in calmer conditions is on the left.
(Barry Hawkins)

During that winter the Smalls were once again returned to darkness. A measure of its necessity was demonstrated at 2 am on 30th December, when the brigantine *Fortitude* bound from London to Liverpool struck, with the loss of 11 hands. The wreckage drifted ashore over the next week. On 18th June 1813 a local newspaper reported the "Lighthouse repaired and light there" heralding the light that once more shone out from the Smalls.

Twenty-one years later, on 20th February 1833, another of these storms was equally destructive in its effect, only this time resulting in a loss of life. A freak wave broke over the

rock, but instead of rolling through the legs, climbed up them to smash its way through the floor of the keepers' quarters, flattening two walls. The iron range in one corner was allegedly compressed flat by the weight of water before being carried out and dropped into the boiling surf. For eight days the petrified keepers had to cook by the lamps in the lantern. One of the keepers later died from the injuries sustained on that day.

By 31st July all was duly patched up, further wooden struts added around the legs, and a new light room returned to service. Despite the many refits it had undergone and the numerous extra legs it had sprouted, Whiteside's structure still managed to retain its unique appearance. It did not, however, appear to be distinctive enough to prevent the *Manuel* of Bilboa, en route from Liverpool to Havana, striking the rock on which it was planted in broad daylight on 9th July 1858. The 12-man crew had just time to leap on to the reef before their badly holed boat drifted off for 3 miles before finally being swallowed by the sea. The keepers fed and kept the crew using the six months' supply of food in the stores for sustenance of its inflated complement, although conditions must have been somewhat cramped.

Although it was certainly isolated, the Smalls rock is in the middle of one of the busiest shipping lanes around our coasts. The majority of trade in and out of Liverpool passed the rock and its lighthouse, as did that bound for Scotland or Ireland. With the 19th century's flourishing sea trade came healthy profits from the light dues, over £6,700 being recorded for one year. Trinity House was well aware of this profitable beacon which was still in private ownership, and in 1823 offered to purchase it. The joint lessees were now the Rev A. B. Buchanan, (Phillips' grandson), and Thomas Clarke, who valued their property at £148,430. The Elder Brethren considered this too high and proceeded no further with the matter. The profits in the meantime continued to mount. By the Act of Parliament of 1836, ownership finally changed hands on 25th March with 54 years of the lease remaining. Trinity House had to pay £170,468 for its purchase, an amount fixed by a local jury as appropriate compensation, yet still £22,000 in excess of their original offer! It is interesting to compare the similarity of these events with those of the Skerries in the previous chapter.

As the 19th century progressed a new breed of lighthouse engineers were erecting their towers on isolated reefs around our shore. The majority of these were of stone construction, giving solid, sturdy towers which resisted the ocean's tumult. Out on the Smalls, Whiteside's wooden light bravely continued in its vigil, although requiring almost constant repair. It was now extremely shaky on its legs, and during heavy seas the whole structure rocked and trembled with each crashing blow, the force of which had been known to fling a clock hung on a wall on to a bunk on the opposite side of the room. It was in 1801 that the great Robert Stevenson had described it, rather unkindly, as, "…a raft of timber rudely put together, the light of which is seen like the glimmering of a single taper." Fifty years later it was still standing, although aged and frail. Few people would care to say how much longer Whiteside's brave wooden beacon would last. It was near to its end, and if Trinity House did not take it down then it would not be long before the Atlantic would reclaim the Smalls.

There were tentative plans for another Smalls lighthouse in 1849 showing that a site had been chosen on the reef for a new stone tower, but it wasn't until 1859 that work commenced. Whiteside's structure had lasted for over 80 years, and while it had needed reinforcement, his principle was basically sound. Today this design is used the world over as a basis for lighthouses in certain situations. This fact ensures that we cannot forget Henry Whiteside and his lighthouse on stilts that he placed on a tiny Welsh rock in the midst of wild seas.

The construction of Trinity House's new tower was entrusted to the capable hands of James Douglass, their chief engineer. He chose the same rock that Whiteside had done, only a short distance away from the existing light. Its profile was based on the now familiar outlines of Smeaton's Eddystone tower – a smooth tapering curve rising gracefully from its base. The designer and consultant engineer was the equally famous James Walker, who paid as much attention to details such as sanitation as he did to the actual construction. In fact, the Smalls was one of the first rock lights to have a water closet incorporated in the design. It was Walker, too, who is credited with the novel idea of building the base of the tower in a series of steps to entrance level, the purpose of these being to break the force of waves crashing against the base and preventing them rolling up the sides. This design was later to be improved upon at such sites as Eddystone and the Bishop Rock by standing the tower on a solid core of masonry. Its function was exactly the same.

While the new tower was under construction, the Smalls was descended upon by a party of Royal Commissioners to inspect progress. The method of construction followed the now standard pattern of cutting and shaping the stone blocks on shore to obtain an exact fit and then transporting them by boat to the site. During their visit the Commissioners noted that:

Each stone has a square hollow on each edge, and a square hole in the centre; when set in its place, a wedge of slate, called a "joggle", fits into the square opening formed by joining the two upper stones. The joint is placed exactly over the centre of the under stone, into which the joggle is wedged before the two upper stones are placed. The result is, that each set of three stones is fastened together by a fourth, which acts as a pin to keep the tiers from sliding on each other. The base of the building is solid. Two iron cranes slide up an iron pillar in the middle, and are fixed by pins at the required position as the work advances. The two are used together, so as to obviate any inequality of strain.

By 7th August 1861, just two years since it was started, the tower was complete. Such a speedy rate of progress was due, no doubt, to the rapidly advancing technology in lighthouse engineering, and also to the fact that the Smalls rock was not continuously being covered by tides or swept by waves, thus hindering the early and most vital stages.

The Smalls now had a splendid new tower, 141 ft high, and a worthy and fitting successor to Whiteside's wooden refuge. At some time subsequent to its inauguration it was painted white with three broad red candy stripes as an aid to identification. Although life in the new tower was considerably more comfortable than in its predecessor, the loneliness and isolation inside its granite walls still continued to test the endurance of the men who inhabited it. One assistant keeper, who lived normally in Ealing where he was a watchmaker, was under no illusion that he "…would rather be anywhere on shore at half the money," because "This is rusting a fellow's life away." Curiously enough, his head keeper, a Welsh farmer, preferred the Smalls to any other post, either a rock station or on shore.

Despite its isolation, modernisation of the Smalls kept pace with the technology of the time. The first example of this occurred 24 years after its completion when the fixed white light that was visible for 15 miles had its character changed to occulting (a steady light with a total eclipse at regular intervals with the period of darkness always less than the period of light) with a fixed red sector light. Two years later the original fog bell was replaced by an explosive signal, originally every 5 minutes but later every 7½ minutes. In 1902 the Smalls was connected to an electric telegraph, but this was, "for life saving purposes only."

Further progress took place in 1907 when a 1st order catadioptric mercury-float optic was installed with the light coming from a 75mm paraffin vapour burner that gave a group flash of 3 every 15 seconds that could be seen from 17 miles away. The optic was rotated by a hand-wound weight-driven clock. At the same time a new red sector light, visible for 13 miles, was introduced in a room below the main lantern, 107ft above the water line, to mark the Hats and Barrels rocks – two satellites of the Smalls reef.

By 1928 the telegraph had been replaced by a telephone connection via a 1¼ins underwater cable which enabled the keepers to report shipping movements to Lloyds. It is also about this time that there are reports of a wind-driven generator being fitted to the lantern roof. There is nothing new, it seems, in wind power. For the next forty or so years things remained essentially the same on the Smalls apart from the arrival of an electric winch, electric domestic lighting and VHF radio telephone.

1970 saw a major modernisation program. The Smalls was the first Trinity House station to be fitted with a 400W metalarc electric lamp, its 1 million candlepower made it visible for 26 miles. This replaced the paraffin vapour burner inside the existing optic. Electricity also took over the rotation of the lantern and powered a newly installed 'racon' (radar-beacon). Down on the reef itself, fuel oil and water storage tanks were constructed about 66 feet from the base of the tower. The fog signal was upgraded to three supertyfon compressed air generators.

In 1978 a helideck of a design almost identical to that on the Eddystone was erected above the lantern to enable helicopter reliefs to be accomplished without the keepers ever leaving the tower. Prior to this date, helicopter reliefs had been established by landing the helicopter on the reef itself adjacent to the tower. These reliefs obviously couldn't take place when the sea was sweeping over the reef or there was a remote possibility of it doing so. Even though this system was theoretically more reliable than a boat relief, it was by no means certain. A helideck above the lantern would make helicopter reliefs virtually guaranteed. Any keeper stationed on this desolate outpost could be almost sure that he would return to his family on time, which must have done much to alleviate the tedium of such a posting. Only fog or a tearing gale might cause a postponement.

The final modernisation on the Smalls reef coincided with Trinity House's ongoing program of automation. The last keepers of the Smalls were withdrawn on 15th May 1986 after a LANBY (Large Automatic Navigation BuoY) was towed into position – delayed by several days because of bad weather, 1½ miles south west of the tower. The automation process involved

large amounts of 'coring' – drilling out 35 metres of circular 'tubes' through the granite walls and floors to accomodate all the new wiring, ducting and piping required to connect up the new equipment. Part of the process also required the installation of a powerful 10kW diesel generator, together with a standby, designed to run for 6 months without attention. Automatic fog detectors set off three powerful new fog signal emitters arranged around the lantern gallery that can be heard from 3 miles away. An emergency beacon in case of a main light failure was installed on the lantern roof below the helideck.

The Smalls lighthouse is still the first sea-mark seen by vessels making for the oil terminals of Milford or the Bristol Channel ports from the west. The broad red and white bands that used to distinguish it from the cold, grey wastes of ocean that surround it were sand blasted off during June 1997, returning it to the natural granite and to an appearance almost identical to when Douglass built it. Its helideck supports were painted red at the same time giving it a vaguely similar appearance to the Eddystone. Since automation and the addition of a ring of solar panels around the helideck the main light flashes continuously – a permanent reminder that although there are no keepers looking out from the Smalls lighthouse, its function to warn of the reef on which it stands is still the primary purpose of the Trinity House service.

Above: A fairly rough day at the Smalls with lots of white water. The lighthouse is now automatic but on the helideck are what look like three rubber bags that would contain fuel for the generators or drinking water. *(Air Images Ltd)*

5 LONGSHIPS
the Land's End light

FEW VISITORS to Cornwall can fail to have been drawn to the very extremity of this country, to the rugged promontory known as Land's End. Here, sea and land meet in conflict; a battleground of crashing waves, foaming surf and precipitous cliffs. Although mistakenly believed to be the westernmost point of the British mainland (it is not, for this distinction goes to Ardnamurchan Point in Argyllshire which extends some 40 miles further west than Land's End), it nevertheless draws countless thousands of tourists each summer to gaze at the last of the English landscape as it drops vertically beneath the Atlantic surf. The ritual of this pilgrimage involves standing on cliff tops to gaze westwards across the grey fields of ocean to the site of the legendary land of Lyonesse, home of Camelot and the Arthurian knights, before it was supposedly engulfed by the sea.

However, their westward gaze will, in all probability, be drawn from the distant horizon to focus on a line of rocks barely a mile from where they stand, on top of which rises the shapely symmetry of the Longships lighthouse. It is a scene remembered, and much photographed by every visitor to this outpost of mainland Britain, more vividly perhaps if viewed at sunset, when the rocks and lighthouse are thrown into silhouette against a flaming sky, to become a lonely convoy of ships that give the reef its name.

The tidal races, swirling mists and half submerged rocks made the Longships a nightmare for vessels rounding the toe of England. A passage between the reef and mainland was fraught with difficulties caused by vicious currents, a confined channel and razor-sharp granite, yet a course too far westwards was liable to draw ships on to the infamous Seven Stones reef. Vessel after vessel was swept to destruction around these few short miles of coast, each met with a gleeful reception from the wreckers who allegedly thrived on the rich pickings

this coastline offered. Nowhere in Britain, it was said, was there a more evil stretch of coast – something the local wrecking fraternity took full advantage of. Children of the coastal villages were taught to pray "God bless father 'n mither an' zend a good ship to shore vore morning," while their parents made considerable efforts to see that their children's prayers were answered by setting up false beacons to lure heavily laden vessels on to a coastline which offered no second chances. It was a lucrative business and one which it was difficult to stop.

Left: When seen against a setting sun from Land's End, the rocks on which the Longships lighthouse was built were thought to resemble a line of ships, hence 'Long -ships'. *(Author)*

As long as this coastline remained unlit, ships would continue to end their days at the hands of the sea, with the likely help of those who gleaned a living from it. For century after century, each night the coast around the Longships and Land's End was plunged into inky blackness where few vessels could survive in heavy seas. John Ruskin once described this area as "…an entire disorder of surges…the whole surface of the sea becomes a dizzy whirl of rushing, writhing, tortured and undirected rage, bounding and crashing in an anarchy of enormous power." The regularity that ships struck this coast forced Trinity House to take notice, although it was not until the end of the 18th century that they were stirred from their lethargy.

It was not solely the Longships which were causing Trinity House concern. In the vicinity of Land's End were two other reefs whose toll of vessels was equally appalling, if not more so, than the Longships. Eight miles distant to the south-west was Wolf Rock, a notorious

Opposite: A stormy day at the Longships lighthouse with white water everywhere. No chance of a boat relief today! *(Author)*

shoal of granite, on to which many ships had piled with catastrophic devastation, and 4 miles to the south-east of Land's End was the Runnelstone, the stumbling block for a good number of vessels attempting the 'short cut' between the English and Bristol Channels through the landward passage past the Longships. If any one of these obstacles were found singly, the navigation around it would have called for the utmost precision, yet having all three together in the same small area of map was a situation which tested the nerves of the most competent mariner. Few ships ventured near these evil rocks during darkness or poor visibility. Tempting fate was one thing, but actually offering the lives of a ship and crew was a risk few captains would take.

In June 1790 Trinity House instructed John Smeaton, after his triumph on the Eddystone, to survey the island of Roseveern in Scilly with a view to placing a beacon there. While this would doubtless aid navigation around the intricate channels of these islands, being nearly 30 miles distant from Land's End it would be of little consequence to the harassed mariners picking their way around the English coast, and so came to nothing. By June of the following year Trinity House had obtained a patent from the Crown for something a little more practical. It was for the erection of a lighthouse on the Wolf Rock, although the Brethren were unwilling to undertake the construction themselves and so leased the rights to a Lieutenant Henry Smith. Smith, too, it appears, was daunted by this task, so the terms of the lease were altered to provide a lighthouse on the more accessible and higher Longships rock, with beacons on the Wolf Rock and Runnelstone. In return for keeping the light dues Smith was charged a rental of £100 for 50 years.

Right: A section through the original Longships lighthouse built by Samuel Wyatt shows the arrangement of the three rooms it contained. This Trinity House plan shows the tower was built on the 'Great Carnvroaz Rock' next to a 'Rock called The Old Man' *(Trinity House)*

A design was prepared by Samual Wyatt, who from 1776 had been architect and consultant engineer to Trinity House. The first stone was laid in September 1791. Cornish granite was used for the blocks, each one undergoing the trenailing and dovetailing which had become accepted practice since Eddystone. They were carefully fitted together in the base at Sennen, then shipped across the 1½ miles of water, before finally being put to rest with waterproof cements on Carn Bras, the highest rock of the Longships reef reaching 40 ft above the waves.

Wyatt's tower was of a short, stumpy design of only three storeys elevation, rising to a mere 38 ft. The walls were 4 ft thick at its base tapering to 3 ft below the lantern and enclosed three rooms, the lowest housed the fresh water tanks, above which was a living room followed by a bedroom. The original lantern was of wood and copper construction and contained 18 parabolic reflectors in two tiers, each with its own Argand lamp. These showed a continuous fixed beam visible for 14 miles. By way of economy, metal sheets were installed against the landward panes of the lantern so no light could be observed from that direction, with a resultant drop in oil consumption. When viewed from the sea Wyatt's tower was not only unusual because of its squat outline, it also possessed stout metal stays which were affixed between the lantern gallery and lantern dome on the outside, suggesting that these might have been added as an afterthought. The first light to end the centuries of darkness around the Longships was seen on 29th September 1795.

With such a powerful indication of the imminent dangers, fatalities amongst the rocks and cliffs fell dramatically. Robert Stevenson, on a visit to the lighthouse in 1801, was much impressed by the light it exhibited, although he was not so enamoured by the conditions in which the keepers lived, especially when he discovered they cooked all their food in the lantern. A subsequent visit 17 years later found this state of affairs had deteriorated further and the whole was "in very indifferent order." The keepers were of a "very ragged and wild-like appearance," and appeared to have been neglecting their duties. Stevenson further noted the poor condition of the reflectors which caused the beam to be "quite red coloured."

Right: A very rare photograph by Nicholas Blake Lobb of Samuel Wyatt's original Longships lighthouse taken in about 1870 before work started on Douglass' second tower. All three keepers are visible; one on the gallery and two on the platform outside the tower.
(courtesy of Mike Millichamp)

In the meantime, Henry Smith was having considerable financial troubles. He was adjudged "incapable of managing the concern" and by 1806 the luckless Smith found himself a resident in the Fleet Prison, London – a well known debtors jail. Being the innovator of the Longships light, Smith was shown exceptional goodwill and generosity by Trinity House who took over the running and maintenance of the light while remitting all profits to his family. It must be mentioned here that the similarity of these events when compared with those that took place at the Skerries lighthouse is remarkable.

Although only a little over a mile from the shore, the existence of the keepers on the Longships could be particularly isolated. For this reason four keepers were assigned to the station, two being on duty at any one time, while the other pair were on shore leave. Spells of duty lasted a month at a time with a month's leave to follow. Wages were £30 per annum with food provided at the lighthouse, although when off duty it was the keeper's responsibility to take care of his own welfare. Many took on second jobs during their time on shore. However, in 1861 an incident is recorded which supposedly caused an alteration to these arrangements. One of the two keepers died, which left his companion to tend the light continuously until he could be relieved. Since then, the story goes, three keepers were stationed in the light at any one time to prevent a similar situation arising. The authenticity of these events is dubious, particularly as they are almost a replica of the occurrences that took place on the Smalls in about 1800.

Such isolation inevitably prompts the imagination of fiction writers with whom the lighthouse is a favourite setting. When this is the case the dividing line between truth and fiction becomes blurred and it can be difficult to separate the fantasy from the fact on which it was probably based. Nowhere can this be better illustrated than on the Longships, where the lighthouse has featured in many different stories of questionable origin, yet all appear to be the result of some particular incident and have a grain of truth.

There exists in a Cornish newspaper of 1873, for instance, the story of a little girl who single-handedly continued to light the lantern every night after her father had been kidnapped by wreckers while fetching supplies on shore. Thinking this would prevent any light being shown from the Longships the wreckers sat and waited for the vessels to blunder their way on to the unlit coast, yet had failed to appreciate the courage of the keeper's daughter who, unable to reach the lantern wicks, had to stand on tip-toes on the family bible to accomplish her task. This story captured the imagination of a contemporary novelist, James Cobb, who in 1878 produced *The Watchers on the Longships* in which this incident appears.

There is also a piece of fiction by the same author concerning a keeper whose hair

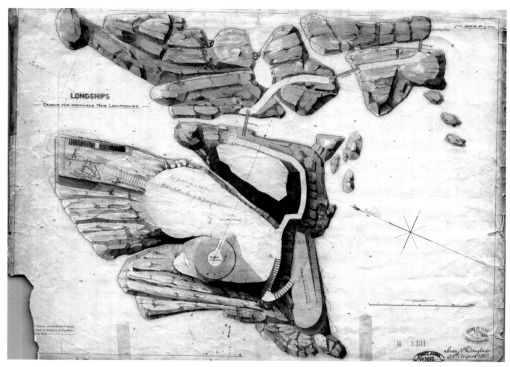

had allegedly turned grey overnight as he had not been informed of the noises generated by air being compressed in fissures of the reef which produced a loud wailing noise when released. This led the author into the erroneous idea that, "more than one untrained keeper has been driven insane from sheer terror." Strangely enough, a newspaper report in December 1842 relates incidents of two years previously when two keepers in the Longships were marooned for an incredible 15 weeks, at the end of which time one of the men, by the name of Clements, arrived on shore with his hair, eyebrows and eyelashes completely grey. He had started his turn of duty with black hair.

During favourable conditions Wyatt's tower proved extremely suitable for the function it was designed to fulfil. Its unceasing beam gave admirable warning of the reef on which it stood, yet during heavy seas its efficiency was seen to drop markedly. Being a little under 80 ft from the high water mark, it was quite within the reach of rogue waves which easily passed over the top of the tower during the many winter storms. Such breakers caused the nature of the light, as seen from ships, to alter from a fixed beam to an occulting flash, ie, a beam whose period of darkness is less than the period of light. The consequences of navigating by such false information could, of course, be disastrous.

Not only this, some of the storms experienced by Wyatt's tower were of such force that the waves had been known to smash the lantern panes and extinguish the lamps. On 7th October 1857 seven burners were doused by water entering the lantern room through the roof cowl, and only a few years later a contemporary writer noted that many of the lantern room windows had been "dashed to pieces by the spray of the ocean during the winter's tempest." This, it seemed, was not an infrequent occurrence, for only 20 years after its construction a report in the *Royal Cornwall Gazette* of 7th January 1815 stated that the Longships lighthouse had been completely destroyed by one of the numerous storms that battered the Cornish coast at that time. It was retracted a week later and a full story carried in the *West Briton* of 13th January. "The violence of the waves broke in part of the roof, which is of thick glass, and for several days the people in the house were unable to ascend to repair the damage. This, however, has now been done and the light is exhibited as usual."

Competent though Wyatt undoubtedly was to hold his position as Trinity House architect, it was obvious that he had seriously underestimated the capabilities of an ocean in fury around the Longships. The water over-topped the lantern with such regularity that during times of heavy seas and high winds the lighthouse, far from aiding navigation through these troubled waters, was a positive liability to it. The solution was clear – a taller structure was required, out of reach of the waves, yet this did not finally materialise until the end of 1873.

Prior to this event came the change of ownership of the old Longships tower from the family of Henry Smith to Trinity House. This was done under the same Act of 1836 which was responsible for the Smalls and Skerries being brought into public ownership. It was also deemed necessary for the same reason – excessive amounts of profit being taken from light dues caused by a boom in sea trade during the early 19th century. Records show that after deducting maintenance monies of £1,183 in 1831 the profit was £3,017. In just five years this had risen to £8,293. Trinity House was now regretting its earlier kindness to Smith and his family who benefited to the tune of £40,676 when the leases were bought back from them in 1836 with only 9½ years left to run.

By 1869 work was all but complete on the Wolf Rock lighthouse. William Douglass, resident engineer at that site, was instructed to prepare plans for a new tower to succeed Wyatt's on the Longships. Supervision of its erection was left in the capable hands of Michael Beazely who had assisted Douglass at the Wolf and followed him from it. In fact, much of the machinery, tools and boats used there were again put to work on the second Longships tower. Granite was the obvious choice of stone, either from Cornish quarries or from northern France.

Block laying commenced in 1870 and being in such an advantageous position, 40 ft above all but the highest seas, work proceeded swiftly. Soon a second tower could be seen rising on Carn Bras. When the tower rose to a suitable height, a passageway was constructed between the two towers which were adjacent to one another. Also, part of the actual reef itself that was in the line of sight from where the entrance door new tower was planned was blasted away. A temporary light fixed to the top of a balance crane that rose from the centre of the new tower took over the warning functions as the new tower grew ever taller.

There were, however, moments of alarm such as took place during the winter of 1872 when 15 men were marooned in the two lighthouses by heavy seas for many weeks. Provisions ran low, despite some being hauled through crashing waves from a supply boat. It was shortly after Christmas that these poor souls could be snatched back into civilisation by a Penzance steamer.

In August 1872 the final stone was laid amid much celebration. The period between then and the end of the following year was occupied with fitting

Left: Another incredibly detailed Trinity House plan dated 30th January 1872 showing how a temporary light was to be placed on top of the growing Douglass tower which was built alongside Wyatt's original light. This plan also shows how much of the actual reef was to be removed from around the new tower to give clear visibility in all directions. (*Trinity House*)

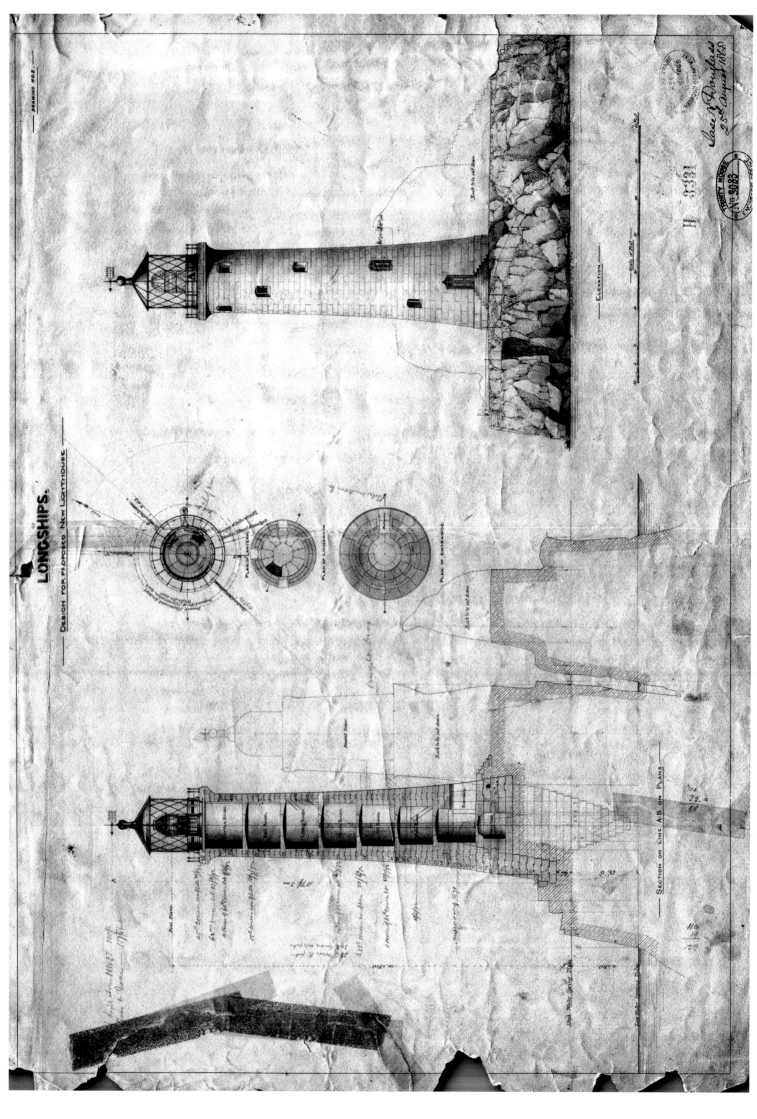

LONGSHIPS.

— DESIGN FOR PROPOSED NEW LIGHTHOUSE —

PLAN OF LANTERN

PLAN OF LIVINGROOM

PLAN OF ENTRANCE

— ELEVATION —

— SECTION ON LINE A.B ON PLANS —

out and the installation of the lantern. Completion was in December 1873 when a fixed white light could once again be seen radiating from the Longships, this time from the top of a 69 ft tower, raising the lantern to almost 114 ft above high water. It had cost £43,870.

The design was very similar to what members of the Douglass family had already produced; a circular granite tower with gracefully sloping sides, the wall thickness diminishing as the height increased, the first few courses of blocks stepped to break the force of the waves, and large fresh water tanks holding 1,800 gallons let into the solid base. A huge gun metal door was positioned at the top of a rather elaborate set of semi-circular steps – the sort that would not be out of place at the front door of an imposing mansion. Behind the door are four rooms; an oil store, living room/ kitchen, bedroom and service room before the lantern itself is reached.

Radiating from the base of the tower are a series of walkways, steps and even iron bridges hewn into different aspects of the reef. They lead to three separate landings called the North Point, the Pollon (Pollard) or South Landing and the Bridges. Having a choice of three enabled stores and keepers to be landed at varying states of wind and tide, although this was no simple task during even the best of conditions. The sea is in constant tumult in and around these granite humps, frothing and boiling angrily as the complex currents and swells collide. A calm day around the Longships is a rare day indeed.

The keepers and their families lived in a row of cottages above Sennen Cove that was visible from the tower. With the use of a telescope it was possible to see the lighthouse clearly from the cottages and so the keepers were able to devise a system of communicating between the two. By painting a white border around the front door it was possible for a keeper to stand on the steps and use semaphore flags to communicate with the shore. Later still came electronic communication. A former principal keeper there once wrote:

"We are rather more fortunate at the Longships than most other rocks, with regard to signalling we being well up in the Morse Code and Semaphore, great strides have been made, we are now able to hold communication with the Wolf Rock, a distance of nearly 8 miles, we also signal to dwellings ashore every night and day and are now able to get all the latest news for the Service and private use which is a very great boon, the old style of signalling by hoisting flags being almost obsolete..."

Right: A 1950s postcard view of Longships lighthouse on what looks like a relief day. The Trinity House flag is flying and two keepers are looking out for an approaching boat from the lantern gallery. It would be another 20 years before Longships got a helideck and boat reliefs became a thing of the past.

Opposite: The original and beautifully hand-painted Trinity House plan of August 1868 for William Douglass' second Longships lighthouse. It has obviously been well used and needed repair work with adhesive tape, but is still a fascinating and detailed document. *(Trinity House)*

Left: An RAF rescue helicopter hovers over Longships lighthouse while something is winched down to the keepers on the gallery. *(Ministry of Defence)*

Wyatt's tower was dismantled soon after the new light was brought into service. It is Douglass' tower which now attracts the wistful gaze of Land's End tourists. During the summer of 1967 modernisation found its way on to the Longships. The old paraffin burners were removed and replaced by a new optical apparatus revolving around an electric bulb. Generators to power this were installed beneath the optics. Replaced at the same time was the gun cotton fog signal by a more modern fog horn. The character of the light has also changed, and now a red and white isophase beam (equal periods of light and dark), every 10 seconds is visible for up to 19 miles in clear weather.

Longships was never a popular posting for Trinity House keepers. It has been described as, "...the most treacherous and dangerous rock in the Service, more lives having been lost there than the whole of our rocks put together..." and, "...it is indeed a very rough station." The comparatively short tower meant conditions were cramped and uncomfortable, and in heavy seas, when all the storm shutters were closed, it was dark and damp inside. Worst of all was the distinct possibility of overdue reliefs.

Unlike most other rock stations the Longships is in the unusual position of being clearly visible in all but the most foul of conditions, yet at the same time highly inaccessible. Before helicopters took on this role, a boat relief of the Longships could be a very uncertain and unpredictable operation. An open boat from Sennen Cove had to edge its way into a narrow gully of Carn Bras to reach one of the three landings – something that was only possible in conditions that were virtually flat calm, and even then the sea between the rocks of the Longships was usually

Left: Communication between the Longships lighthouse and the keepers cottages at Sennen used to be done by semaphore – as demonstrated here by two of the off duty keepers and their very disinterested dog. *(Mike Millichamp)*

anything but. If the boat couldn't tie up at the landing it was sometimes possible to transfer keepers and stores on the end of a jib – but often at the expense of a soaking.

Ever since Carn Bras has had a lighthouse on it these landings have suffered much torment from heavy seas and swells which frequently wash over and smash rocks against them, even in comparatively calm conditions. Centuries of storms have broken and shattered the rock, requiring much repair work for the landings to remain serviceable. The last major damage occurred in December 1990 when violent storms smashed a huge chunk off the Pollard landing requiring considerable remedial work with rock bolts to prevent further deterioration. Delayed boat reliefs were part and parcel of life on the Longships.

Things improved no end when a helideck was erected above the lantern in the late 1970s. The helicopter journey from the Trinity House cottages at Sennen took just two minutes and was independent of sea conditions, unless these were so bad there was a possibility of waves reaching the helideck itself – not uncommon during winter storms. The helicopter relief certainly improved travel to and from the Longships, but it couldn't improve conditions within the tower for the keeper's month of duty.

It was a prime candidate for automation and the inevitable started in 1987 and was

Opposite: Westerley gales can drive the waves to spectacular heights up, and sometimes over, the Longships lighthouse. Each wave causes the tower to shudder and the whole of the Longships reef becomes a confused scene of white water. *(Paul Campbell)*

completed early in 1988. It followed the now standard procedure of installing new diesel generators, electric fog signals and a fog detector. The fresh water tanks in the base of the tower now store enough fuel oil for the generators to operate for 6 months without attention. All this was done without interrupting the normal operation and warnings the lighthouse gives. It is now monitored and controlled, like all the rest of their lighthouses, from the Trinity House Control Centre in Harwich.

Right: The Longships' tower casts a long shadow on an autumn afternoon, while on the helideck we can just see a silhouette of a keeper lowering supplies through the trapdoor.
(Author)

Opposite: The three keepers of the Longships lighthouse anxiously watch the approach of the Trinity House helicopter in 1982. One or two of them will be leaving in it for a month's leave.
(Author)

6 THE LONGSTONE
Grace Darling's light

NO WORK about isolated lighthouses or lighthouse stories would be complete without containing what is probably the most well known story about a British lighthouse ever recounted. It is set off the rocky Northumberland coast at a lighthouse known as Longstone, and while this name might not be immediately recognisable, the name of the lightkeeper in 1838 will be instantly familiar to generations of school children to whom this drama has been told. His name was Darling, and it was the courage of his daughter Grace which has ensured the place of the Longstone lighthouse in our maritime history.

The Longstone lighthouse was built on the outermost of a series of about 25 islands known collectively as the Farne Islands, situated some 7 miles off the Northumberland coast – a favourite perch and breeding site of countless thousands of guillemots, puffins, razorbills, terns and gulls. The story of the Farne lighthouses is a long and involved series of events which began well over a century before Grace Darling was to focus attention upon them. In essence, the lighting of these dangerous islands and reefs was accomplished in a rather haphazard manner, with the construction, modification, abandonment or demolition of several towers, of varying effectiveness, placed on several of the islands within the group. The present Longstone tower is a comparatively recent addition to the Farnes.

The isolated nature of the islands made them a favourite refuge of hermits, recluses and saints. It is said, for instance, that St Cuthbert lived alone on Farne in the 7th century. Although only a handful of miles from the mainland, their isolation was due in most part to the fact that their rocky nature and low profile – many rising only a few feet out of the surf – made them difficult to approach and an even greater danger to shipping wishing to avoid the half-submerged reefs and craggy ledges. During the 18th and 19th centuries the inhabitants of the Farnes were known to supplement their income from the salvaging, or looting, of wrecked vessels, much as their Cornish counterparts were prone to do. A premium was paid, which doubled after midnight, to the first person reporting a shipwreck, while the Sailing Directions for the early 19th century gave little comfort by noting that, "Dead bodies cast on shore are decently buried gratis."

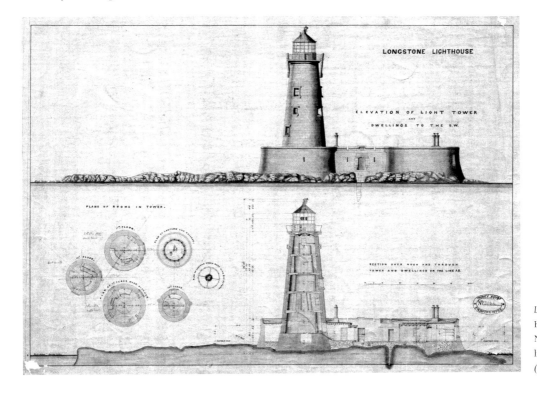

Right: A painting by Thomas Musgrove Joy showing Grace Darling and her father in an open boat trying to rescue the survivors from the wreck of the *Forfarshire* on the Big Harker rock. In the background is the Longstone lighthouse. *(RNLI)*

Right: The scene in the kitchen at the Longstone lighthouse after the rescue of the survivors from the *Forfarshire*. Grace and her father are on the right, comforting the survivor wrapped in a blanket, while her mother offers hot soup to another. *(RNLI)*

The history of the Farne Islands lighthouses starts in 1669 when Sir John Clayton, a long-time antagonist of Trinity House, applied for a patent to erect four lighthouses along the east coast of England at the Farnes, Flamborough Head, Foulness near Cromer, and Corton near Lowestoft. He was granted the patent for all the proposed sites and proceeded to erect stone towers at all four locations at a cost of £3,000. However, although the Crown had specified dues of 2½d per ton for laden vessels and 1d per ton for unladen ones, payment by merchants would be voluntary. He lit the tower at Corton on 22nd September 1675 but because of difficulties in collecting dues, not helped by Trinity House establishing a rival light for which they charged no dues, he was unable to light the other three towers and eventually had to extinguish the light at Corton in 1678.

Clayton's good intentions had left the Farne Islands with a stone tower with no fire on it, yet the islands still claimed shipping with alarming regularity. Representations were made by local shipowners to Trinity House who successfully applied for a patent, and in due course leased this on 6th July 1776 to Captain John Blackett and his family, who lived on the islands. He in turn built two lighthouses, one on the Inner Farne Island and the other on the smaller Staples or Pinnacle Island. On the Inner Farne he erected a 40 ft stone tower, known as Cuthbert's tower, on top of which was lit a coal fire, while on Staples Island a stone cottage was built with a sloping roof and a glazed lantern placed centrally on the ridge. The light source for this lantern was thought to be oil lamps. It is at this cottage that the first connection between the Darling family and the Farne Islands can be seen, for Robert Darling, the grandfather of Grace, was a lightkeeper here in 1795.

It was not long before complaints were heard regarding the positioning and effectiveness of the Staples Island light. To improve matters a further roughly-built tower was erected in 1791 on the adjacent Brownsman Island which supported a coal-burning grate. This was approximately 21 ft square in section and about 30 ft high with an external ladder.

Left: The actual boat in which Grace Darling and her father carried out the rescue of September 1838. It has been named after her and now rests in the Grace Darling Museum, Bamburgh. (*RNLI*)

Even this measure did little to improve the visibility of the lights which could frequently be obscured by smoke. Trinity House was naturally keen to replace the inefficient coal grates with Argand lamps, and so with a new patent of November 1810 proceeded to build two further lighthouses, under the supervision of Daniel Alexander the Trinity House architect, one on the Inner Farne adjacent to the existing tower, and the other on the Longstone rock – an outlier of the group. Both lanterns were fitted with Argand lamps and reflectors and had revolving optics. They cost together £8,500.

Trinity House granted a further lease to the Blackett family which gave them the profits from the two lights after maintenance and running costs had been deducted by the Corporation who had now taken over the running of the two stations. This was yet another example of generosity by Trinity House which they were later to regret and practically a carbon copy of the events at the Skerries lighthouse dealt with earlier. When the Elder Brethren were keen to buy back the lease of this lucrative source of funds in December 1824 with 15 years of the lease still to run, it cost them a staggering £36,445 for the two lighthouses. Yet this was not the final cost. It had been decided that a new stone tower was required on the Longstone reef, to be built by Joseph Nelson, which in turn was to cost a further £6,063.

The new Longstone tower was 85 ft from base to vane and painted red with a white band to make it an easily distinguishable daymark. It had a lightkeepers' house attached and was enclosed by a boundary wall. When its revolving white flash every 30 seconds was first seen on 15th February 1826 it was the signal for the demise and subsequent demolition of the tower on Brownsman Island. The keeper was transferred to the Longstone light – his name was William Darling and he took with him his wife and 11-year-old daughter Grace, shortly to become more famous than William could ever have dreamt possible.

Grace Horsley Darling was born in the year of the Battle of Waterloo on 24th November 1815 at Bamburgh, Northumberland. Her early years were spent with her parents, brothers and sisters in the lighthouse on Brownsman Island. Her education, apart from a basic grounding in elementary subjects, was what she learnt from her parents and she soon developed characteristics that were clearly inherited from her father – untold patience, conscientiousness and steadfastness. She was a quiet, unassuming, well principled girl who gave no hint of having the slightest heroic streak within her. After her brothers departed to continue their education on the mainland, Grace was left to continue her life helping her father with his duties in the new Longstone lighthouse.

Having such an intimate relationship with the sea, Grace would have witnessed all its moods; a placid millpond in which the Farnes' seal colonies gleaned their food, then quickly transformed into a raging killer as an 8-knot tide ripped between the rocky ledges. Mountainous seas often crashed on to sea-polished islets such as the Big and Little Harkers, Nameless Rock, or Gun Rock. She would have seen that the rock on which the lighthouse stood, only 4 ft above the high water, was swept in every gale by curtains of spray and foam, a scene that would fix in her memory the fact that the Farnes offered no half chances to vessels unlucky enough to drag their hulls over the teeth which surrounded her home.

In 1838 Grace was 22 years old and her life had so far been devoid of the sort of drama which would shortly bring her to public prominence. On 5th September, the *Forfarshire*, a luxurious new pleasure steamer, left Hull on one of her regular trips to Dundee with 63 passengers. During the night of 6th–7th the wind veered to the north and rose in strength to gale force with an accompanying worsening of sea conditions. On Longstone, William Darling and Grace busied themselves with securing the lighthouse's coble boat to its supports and

Left: An interesting view that shows how close the Farne Islands are to the Northumberland coast beyond. The extensive rock above the lighthouse tower is the Big Harker from which Grace Darling and her father rescued the survivors of the *Forfarshire* in 1838.
(*Author*)

generally preparing themselves for the gale which, as fate would have it, coincided with a high tide. They knew to expect a rough night but were not unduly perturbed.

On board the *Forfarshire* steaming steadily northwards along the east coast, a problem with the boilers had become apparent. A leak had sprung allowing steam and boiling water to escape into the engine room. Captain Humble ordered the pumps to be started in an effort

Right: One of the many portraits of Grace Darling painted after her heroic rescue. *(RNLI)*

to clear the water but these were unable to deal with the vast amounts of scalding water now escaping from the boiler. Eventually, the sheer volume of water extinguished the boiler fires and the ship was left without power in a rising gale. With storm force winds beating down from the north on the crippled vessel, the captain gave orders for the sails to be set in a desperate attempt to keep the ship on course. It was a futile gesture in such conditions, the ship went about and became unmanageable, and efforts to effect some control by dropping the anchors were unsuccessful.

Captain Humble was in a desperate situation. He was in command of a vessel which had very little manoeuvrability and no engines, but it carried over 60 passengers and a valuable cargo. He had little option therefore but to try to find shelter amongst the Farne Islands, now appearing off the port bow, as it would clearly be folly to try to ride out the storm on the open sea. The light from the Longstone lighthouse shone bright and clear through the spray, but it was about this time that Captain Humble made the mistake that was to cost him the life of his ship and many of its passengers. He somehow mistook the position of this light, or was even confused as to which light he could see, and with his vessel drifting helplessly south on the tide the *Forfarshire* struck, between three and four in the morning an outlier of Longstone rock called the Big Harker, a rock rising sheer from 600 ft below the waves and so sharp and rugged that standing upright on it was said to be impossible, even when its surface was dry.

Left: Another Grace Darling portrait painted by Henry Parker on 10th November 1838 at the lighthouse. *(RNLI)*

Almost immediately the vessel was ripped open and she began to break up. Most of the cabin passengers and crew were washed off the decks by mountainous seas, never to be seen again. A few lucky souls escaped in a boat. Captain Humble, with his wife in his arms, were last seen being swept out to sea. About 13 souls were fortunate enough to be able to scramble on to the rock or cling with numbed hands to pieces of broken vessel wedged in the crevices of the Big Harker. A constant deluge from curtains of icy water threatened to drag them from their exposed perch back into the foaming waters around them. Several hours of such an ordeal were too great for two adults and two children who were snatched back by the sea when exhaustion prised loose their grip.

Inside the Longstones lighthouse Grace was having difficulty in sleeping. She was awoken just before dawn by the shrieking of the wind, but above this noise she thought she heard other cries – shouts of distress and anguish produced by human voices. She roused her father who produced a telescope and peered into the half light towards the direction of the noise. He was shocked by what he saw. Half a mile away he could just pick out the shattered remains of a once proud vessel, her masts snapped and sails shredded, being pounded against the Big Harker rock by heaving seas; but worse still was the small group of miserable souls huddled together on the adjacent rock in fear of their lives.

Left: William Darling, Grace's father *(RNLI)*

Both Grace and her father knew that no boat would be able to put out from Bamburgh in such conditions and that if the survivors were not to perish a rescue would have to be attempted from the lighthouse. Mrs Darling was loath to let her husband and daughter undertake such a rescue, fearing that the wild seas breaking around the rocks would claim her husband and daughter also. No such thoughts entered Grace's mind – she launched the flimsy coble that was kept at the lighthouse and with her father and her rowing in unison, set out across the raging ocean toward the shipwreck. The survivors were less than half a mile from the Longstone yet the sea forced the two rescuers to row for over a mile through the heaving swell and swirling currents before they were able to navigate close enough to the rock with the luckless survivors on. The flat-bottomed coble was not large enough to hold all the survivors so William Darling leapt on to the rock and assisted into the frail boat a woman survivor, who had earlier watched both her children swept to their deaths, one injured man, plus three of the crew. Grace in the meantime was fighting with the oars in a desperate attempt to hold her boat close to the jagged rocks, a feat of courage and strength of the highest order – the slightest misjudgement would have reduced the coble to splintered matchwood.

Right: The rescue of the survivors of the *Forfarshire* wreck is commemorated by this wooden plaque on the wall of the room that was Grace Darling's bedroom. *(Author)*

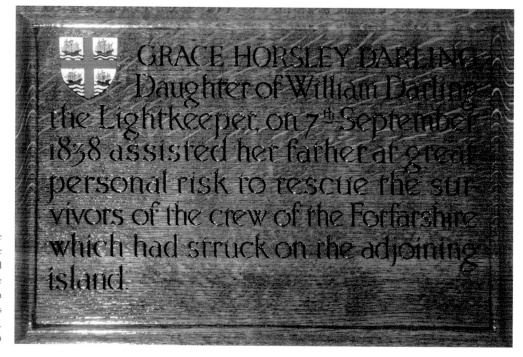

GRACE HORSLEY DARLING Daughter of William Darling the Lightkeeper, on 7th September 1838 assisted her father at great personal risk to rescue the survivors of the crew of the Forfarshire which had struck on the adjoining island.

Grace and her father returned to the lighthouse with the five survivors where her mother was waiting to help the injured man and woman into the shelter of the light. Grace remained with them to tend to their injuries while her father and two of the male survivors returned again to fetch off the remaining passengers without further loss of life. Half an hour after the rescue, much to the surprise of William Darling, a boat could be seen approaching the scene of the wreck. It contained the coxswain of the North Sunderland lifeboat, William Robson, his two brothers, and also Grace's brother, William Brooks Darling. They had managed to put out from Seahouses but were too late to offer any effective help. The survivors spent three days in the lighthouse before they could be taken ashore to tell the story of Grace Darling and her heroic deeds.

Even so, it was many days before Grace's part in the rescue became widely known. It was only when *The Times* asked "Is there in the whole field of history, or of fiction, even one instance of female heroism to compare for one moment with this?" a fortnight after the event, that more and more attention was focused on Grace Darling who was slowly but surely becoming Britain's first national heroine. The public was clamouring to know more about this frail little girl and her act of bravery.

Longstone Lighthouse 12876

Left: Probably a 1930s scene at Longstone lighthouse with a flotilla of open boats wedged into one of the creeks of the rock. The diaphone fog horn projecting from the roof of the building was bombed during World War II and never replaced – but the tower survived. It is difficult to tell from this view whether the lighthouse had acquired its white stripe yet.

Countless songs and verses of poetry were written of her deeds, even William Wordsworth was moved enough to produce his own tribute containing the lines:

> *And through the sea's tremendous trough*
> *The father and the girl rode off.*

She was the recipient of bravery awards from many sources, including a gold medal from the British Humane Society, as well as sacks full of letters that were somehow delivered to Longstone. A national fund was raised with considerable sums of money being kept in trust for the somewhat bemused Grace. She and her father had to sit for seven artists within 12 days while others waited their turn. Locks of hair, supposedly Grace's, and squares of material allegedly taken from the gown in which the rescue was undertaken were sold in vast quantities throughout the country. Steamers ran excursions to the Longstone lighthouse, and Grace even had offers to appear on the stage. Flattering testimonials were bestowed upon her, yet through all the national obsession with Grace Darling and the wreck of the *Forfarshire*, Grace herself remained comparatively unaffected by all the commotion, continuing to assist her father with his duties as she had done in the past.

Indeed it would appear that her father did not view the events of the rescue with such a sense of drama as did the rest of the country. It is clear from his entry in the lighthouse journal that such events were regarded as part and parcel of a lightkeeper's life – the rescue was only one aspect of his function to save lives wherever humanly possible, and Grace's involvement was coincidental in so much as she was only fulfilling the role of her brother who would normally have assisted his father had he not been on the mainland. The relevant entry in the log reads:

The steam-boat Forfarshire, *400 tons, sailed from Hull for Dundee on the 6th at midnight. When off Berwick, her boilers became so leaky as to render the engines useless. Captain Humble then bore away from Shields; blowing strong gale, north, with thick fog. About 4 am on the 7th the vessel struck the west point of Harker's rock, and in 15 minutes broke through by the paddle axe, and drowned 43 persons; nine having previously left in their own boat, and were picked up by a Montrose vessel, and carried to Shields, and nine others held on by the wreck and were rescued by the Darlings. The cargo consisted of superfine cloths, hard-ware, soap, boiler-plate and spinning gear.*

Nowhere, it will be noted, does Grace's name appear, in this or any other entry relevant to the rescue. This was not out of spite towards his daughter; on the contrary, Grace was often regarded as his favourite child, but purely because it was considered perfectly normal for his daughter to assist him with his duties in the absence of his son, whether these duties be the mundane trimming of the lamp wicks or something a little more dramatic. Very little is mentioned in the lighthouse journal of Grace's national fame apart from short entries regarding artists that visited the lighthouse followed by meticulous detail of the fish he caught or unseasonal breeding patterns of the sea birds. Clearly, William Darling thought very little of all the attention the rest of the country were paying his family.

In 1841, just three years after the *Forfarshire* rescue, Grace, who had a strong frame but was never particularly robust, became ill to such an extent that it was clear her health was deteriorating. What was at first diagnosed as a common cold did not respond to treatment and a subsequent investigation confirmed that Grace was suffering from tuberculosis. She was taken to Bamburgh for further treatment and appeared to make some improvements. Sadly, it did not last, and she died on 24th October 1842 at the age of 26. She was buried in Bamburgh churchyard where a monumental tomb was erected in her memory in 1846. On this her effigy is in a recumbent position, hands crossed with an oar by her side embraced by her right arm. Inscribed in the stone is another tribute from the pen of William Wordsworth:

> *The maiden gently, yet at Duty's call*
> *Firm and unflinching as the lighthouse reared*
> *On the island-rock, her lonely dwelling place;*
> *Or like the invincible rock itself that braves,*

Age after age, the hostile elements,
As when it guarded holy Cuthbert's cell.

Above: It is 1952 and modernisation is well under way at Longstone. Scaffolding, cranes, piles of rubble and workman's huts all suggest there is much work being done at the lighthouse.
(ALK archives)

In 1859 a party of Royal Commissioners visited the Longstone lighthouse which they found in excellent order. William Darling again recounted the story of the rescue, pointing out the Harker rock from the tower. He told them that his daughter had died of a decline brought on by an "anxiety of mind" because "so many ladies and gentlemen came to see her that she got no rest."

Grace Darling had become a national heroine almost overnight. Sadly, it was perhaps because of her deeds of courage and bravery, because of her skill of seamanship, that she suffered such an untimely and premature death. However, such events are now so well documented that her memory will live on, helped by the more tangible remains of the coble that she and her father used which is on display in the Grace Darling museum, while a plaque on the wall of Bamburgh post office reminds visitors that this is the house in which she died.

7 BELL ROCK

the start of the Stevensons

IF WE ARE to look for a reef which has a comparable history of disaster to, say, the infamous Eddystone, then it is surely to be found off the east coast of Scotland, where the Bell or Inchcape Rock rears its extensive bulk. The history behind this wicked reef is as famous as the lighthouse which now stands upon it, a history which stretches back to the 14th century when John Gedy, The Abbot of Aberbrothock (the ancient name for Arbroath) allegedly fixed upon it a bell tolled by wave action as a gesture to prevent further destruction and loss of life. The bell was removed by a Dutch sea-pirate who later perished on that very reef after losing his way in a storm. Since then the reef has been known the world over as Bell Rock.

During the 16th century this legend was perpetuated by a Scots historian who wrote, "On this great hidden rock...there was a bell fixed upon a tree or timber which rang continually, being moved by the sea, giving notice to the saylers of the danger. This bell or clocke was put there and maintained by the Abbot of Arbroath and being taken down by a sea pirate, a yeare thereafter, he perished upon the same rocke with ships and goodes, in the righteous judgment of God."

In 1815 the poet Robert Southey ensured the immortality of these events when he wrote the ballad of 'Sir Ralph the Rover' in which he describes:

> The pious Abbot of Aberbrothock
> Had placed that bell on the Inchcape Rock;
> On the waves of the storm it floated and swung,
> And louder and louder its warning rung.
>
> When the rock was hid by tempest swell.
> The mariners heard the warning bell;
> And then they knew the perilous rock,
> And blessed the Abbot of Aberbrothock.

Sir Ralph the Rover was Southey's name for the pirate who upset the kindly Abbot's intentions by severing the bell from its float,

> Down sank the bell with a gurgling sound,
> The bubbles rose and burst around;
> Quoth he, 'Who next comes to the rock
> Won't bless the Abbot of Aberbrothock'.

On return from one of his plundering raid, heavy seas and thick mists upset his navigation:

> 'Canst hear,' said one, 'the breakers roar?
> For yonder, methinks, should be the shore.
> Now where we are I cannot tell –
> I wish we heard the Inchcape Bell!'
> They hear no sound – the swell is strong;
> Though the wind hath fallen they drift along,
> Till the vessel strikes with a shivering shock –
> 'O Heaven! it is the Inchcape Rock!'

Opposite: The Bell Rock lighthouse with the walkway just visible between the lighthouse and the helicopter landing platform before it is finally engulfed by a rapidly rising tide. *(P. D. Green)*

Sir Ralph the Rover tore his hair,
And cursed himself in his despair.
The waves rush in on every side;
The ship sinks fast beneath the tide!

Down, down they sink in watery graves,
The masts are his beneath the waves!
Sir Ralph, while waters rush around,
Hears still an awful, dismal sound –

For even in his dying fear
That dreadful sound assails his ear,
As if below, with the Inchcape Bell,
The devil rang his funeral knell.

It is doubtful whether any bell existed or such events actually took place, although the regularity with which vessels continued to strike the Bell Rock and perish was a figment of no one's imagination. During a particularly severe three-day gale in December 1799, over 70 ships foundered on Scottish coasts, many of them had been driven northwards from England. At least two of these met their end on the Bell Rock including, it is said, HMS *York*, a warship of 74 guns which was lost with all hands.

The Commissioners of Northern Lighthouses had been entrusted with the erection and maintenance of beacons in Scottish waters since 1786. They were not blind to their responsibilities over Bell Rock but could find no bank willing to advance the many thousands of pounds which would be necessary for the placing of a beacon on such a hazardous site. Its design, too, was proving elusive. Many and varied were the structures proposed by private individuals – stone towers on hollow metal legs filled with sea water, stone towers on stone legs, a solid stone base built on a wooden raft, floated into position and sunk. There were more,

Right: This is what the Bell Rock reef looks like as a rising tide with a strong wind stirs the whole reef into a maelstrom of white water. There were no second chances for vessels driven onto this particular graveyard.
(Jim Bain)

but most were pure fantasy on behalf of their designers. In the early years of the 19th century three wooden beacons were actually placed on the rock but were swiftly swept into oblivion by the waves.

Left: At high water there is no sign of the Bell Rock reef – it appears as if the lighthouse rises straight out of the sea. *(Eddie Dishon)*

Bell Rock continued to claim its victims to such a great extent that the Northern Lighthouse Board was forced to act. In 1800 the members detailed their chief engineer, Robert Stevenson, to study the feasibility of erecting a warning structure on this "frightful bar to navigation." He made extensive tours around the existing lighthouses of the British Isles before arriving at the conclusion that for such a site only a stone tower would be resilient enough to withstand the elements. A design was needed, Stevenson said, along similar lines to Smeaton's structure on the Eddystone. Having said that, he also realised that despite there being an existing blueprint to work from, a similar construction on the Bell Rock would present exceptional technical difficulties which had never before been encountered on a wave-swept site. For the reasons behind this we must look at the actual Bell Rock in more detail.

The story of this lighthouse has many parallels with the Eddystone, although the reef on which it stands could not be more different. Whereas Eddystone appears as three separate, jagged and converging ridges, rearing sharply from the English Channel, the Inchcape presents itself as a broad, serrated bed of sandstone, rising 27 miles east of Dundee and some 11 miles distant from Arbroath. Its area is extensive when compared with other sea rocks, being approximately 2,000 ft long by 330 ft wide, rising sharply on three sides but shelving away

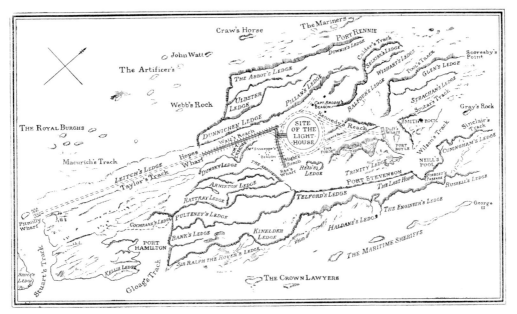

Left: One of Stevenson's early plans of the Bell Rock showing the site of his proposed lighthouse, the cast-iron tramways he proposed to build, and the site of his beacon. Many of the names given to the ledges, tracks and channels were those of the workman building the light.

gently towards the south. Its position, straggling the eastern and northern approaches to the Firths of Forth and Tay made it a fearful obstacle for coastal traffic, to be avoided at all costs. Many vessels did not and paid the price.

However, if these are not facts enough to convince even the most sceptical person of the malice that lurks here, there is one more feature of this reef which makes its other characteristics pale into insignificance; one fact that made this reef the most evil spot along the eastern seaboard of the British Isles, and a challenge never before encountered by any

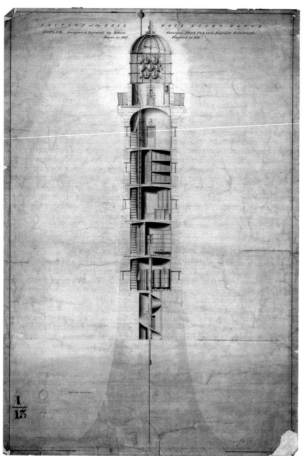

Right: A plan of the lighthouse that has been delicately hand painted and shaded to show what it would have looked like immediately after completion. Details such as the clock in the service room below the lantern or what looks like a marble bust above a window in the room below make this drawing a real work of art.
(Northern Lighthouse Board)

lighthouse engineer throughout the world. It is simply that at high water the Bell Rock is submerged by 12 to 16 ft of water, making it completely invisible. At low water a scant 4 ft appears, and between tides this whole weed-covered submarine plateau becomes a confused scene of white water and swirling currents.

If any structure was to be built here the workings would be completely immersed twice a day until they could rise above the high water mark. The time actually possible on the rock for construction would be scarce indeed before the rising waters drove the workmen from their site. Smeaton's Eddystone, set upon a site above high water at all times, was considered an unsurpassable achievement, yet here was Robert Stevenson attempting to emulate him on a rock immersed twice daily by the notorious North Sea. Stevenson knew that this work would test all his powers as an engineer.

It would also be a costly venture. Smeaton's stone tower had cost somewhere in the region of £20,000. Stevenson estimated that about £43,000 would be necessary to finance the erection of a similar structure on the Bell Rock. The Northern Lighthouse Board did not have such money in their possession but an Act of July 1806 provided for a levy to be extracted from vessels passing the rock, and an advance of £25,000 was paid for the venture to be started.

In 1807 work on construction began. John Rennie, probably the leading civil engineer of the day, was appointed as consulting engineer by the Lighthouse Board, to advise and assist Stevenson in his task, although, as we shall see later, the actual design and day to day routine of construction was very much the work of Stevenson, despite a claim by Rennie's son to the contrary.

During the early months of 1807 the tedious but necessary business of employing masons, blacksmiths and foremen began. A base was established at Arbroath where the stones would be cut and dressed before despatch to the rock. Two sailing vessels were acquired by Stevenson. The *Pharos* was a 70 ft Prussian fishing vessel which was to serve as a temporary light and barrack for 30 of the artificers, while the *Smeaton* was a specially built tender of 40 tons to be used for the ferrying of rough stone from the quarries into the yards at Arbroath, and the finished product to the reef. She could also accommodate 24 men. The *Pharos* was towed into position in July and anchored a mile off the rock, from where a light was exhibited on 15th September onwards. On 17th August the *Smeaton* was sailed to her moorings a quarter of a

Left: Robert Stevenson (1772–1850) the engineer of the Bell Rock lighthouse.
(Northern Lighthouse Board)

mile from the rock, while Stevenson and six masons landed to select a site for the lighthouse and wooden barrack they planned to erect as a refuge for the men while on the reef.

It was seven years since Stevenson had first set foot on the reef at the request of the Commissioners. On that day in 1800 he, and the boatmen who had brought him there, were vividly reminded of the reef's morbid record. Scattered over the surface and lodged in its crevices was the debris from the wrecks of centuries past, each a catalogue of human misery and despair. Stevenson described the activities of his companions on that first visit:

...meantime, the boatmen were busily employed in searching all the holes and crevices in quest of articles of shipwreck, and by the time the tide overflowed the rock, they had collected upwards of 2 cwt of old metal, consisting of such things as are used on shipboard...such as a hinge and lock of a door, a ship's marking iron, a piece of ship's caboose... a soldier's bayonet, a cannon ball, several pieces of money, a shoe buckle, &c,...a piece of kedge anchor, cabin stove, crowbars, &c.

Once the site for the wooden refuge had been selected work could be commenced on the 12 holes, 2 ins in diameter and 20 ins deep, which would hold it securely to the reef. The sandstone was resistant to the blows of their tools which blunted quickly. The comparative spaciousness of the rock allowed a portable forge to be set up which returned the men's tools to their former keenness, although not without some considerable skill on behalf of the blacksmith who had to maintain the forge while being continually harassed by breaking waves and tides which rose with alarming speed.

During this preliminary work a much-documented incident occurred on 2nd September. There were 32 men working on the rock, landed from three boats. Work proceeded until the winds freshened and the sea took on a mood of turmoil. Two men left the reef in one of the boats in an attempt to secure the moorings of the *Smeaton*, a fruitless mission as it turned out for the mounting seas had broken her ropes and she was drifting seaward at a rate of knots, dragging

the small boat with her. Stevenson happened to look up and noted the situation. It was quickly apparent that this predicament was somewhat perilous. Stevenson wrote later that he found himself:

...placed between hope and despair, but certainly the latter was by much the predominant feeling in my mind – situated upon a sunken rock in the middle of the ocean, which, in the progress of the flood tide, was to be laid under water to a depth of at least 12 feet in a stormy sea. There were this morning 32 persons in all upon the rock, with only two boats, whose complements, even in good weather, did not exceed 24 sitters.

The *Smeaton* was by now 3 miles distant when the rapidly rising tide called a halt to the work. The men, still oblivious of the circumstances, made their way to where the boats were moored to find that only two remained. Two boats which normally only carried eight men in such rough conditions. The men stood in silence. Stevenson was lost for words. Their plight was all too obvious and needed no verbal explanations. Stevenson had decided that the only solution to what appeared to be an insoluble problem would be for some men to take to the boats while the remainder hung from the gunwhales.

By the grace of God, before he actually presented his plans, a boat was spotted approaching the reef. It was the pilot boat from Arbroath delivering the mail. Only a stroke of good fortune brought the boat close, for seeing the *Smeaton* disappearing into the distance the captain assumed that all hands would be on board and was about to return to port. It was only at the last minute the stranded men were seen. Sixteen souls were plucked from the reef by the pilot while the remainder scrambled into the boats. Three miserable, storm-tossed hours passed before they made contact with the *Smeaton* close to midnight, to bring to an end the day many thought would be their last. So shaken were the men that only eight would return to the reef the following day. The safe return of this gallant party restored confidence in the others who returned to normal working.

By sheer coincidence, a similar incident is reported just three days later on 5th September, when the *Pharos* had her moorings cut by a piece of floating wreckage. Luckily there were no men on the reef when this occurred and the *Pharos* was returned to a mooring slightly more distant from the rock.

Throughout the remaining days of September large numbers of men, sometimes as many as 52, toiled on the reef to erect the wooden struts for the refuge beacon and secure them into the

Opposite: A striking image taken by a former keeper at Bell Rock from the helipad as the white water roars in on a rising tide.
(Jim Bain)

Left: Before automation and helicopters, the reliefs at Bell Rock were done from open boats which tied up at one of the cast-iron walkways that radiate from the lighthouse. Everything then had to be hauled up through the entrance door.
(Northern Lighthouse Board)

Left: Thirty years on, a picture from almost exactly the same spot. This time the relief is done by helicopter, but everything landed still had to be hauled through the entrance door.
(Author)

rock by means of metal clamps. Twelve wooden beams were fixed together to give a pyramidal structure whose base diameter spanned 36 ft, which gave a height of 55 ft above the reef. Cross members and raking struts gave extra stability and the whole was further strengthened by chaining it into the rock. The first 12 ft of the legs were coated with pitch to resist saturation by sea water, and lashed to the very top were numerous cases containing 5 lb of biscuits and 48 bottles of drinking water for use by any unfortunate mariners wrecked in the vicinity. With the framework of the barrack in place, and five days spent in excavating the base of the tower, work finished for the season during October.

In the winter months, the now familiar routine of preparing and shaping the huge granite and sandstone blocks using a template was followed. Stone was ferried into the workyard from two sites; sandstone from Mylnefield near Dundee, and granite from Rubeslaw at Aberdeen. At Arbroath gangs of men laboured meticulously to dovetail the blocks and cut holes through them for trenails.

A new tender, *Sir Joseph Banks*, was constructed, and that in turn carried three new rowing boats for the speedier conveyance of the prepared blocks, as well as incorporating living quarters for 15 men. It was during this winter that plans were also in hand for a novel idea designed to overcome the problem of moving the blocks, which could weigh as much as 3 tons, from the point

Left: This is a really unusual view of Bell Rock lighthouse – the top of the lantern dome. A couple of keepers work on the emergency beacon while another captures the scene from the ladder that leads to the anemometer – the highest point it is possible to climb to on the lighthouse. *(Jim Bain)*

of off-loading from the boats to the site of the tower. In certain circumstances this could be almost 300 ft across a rock surface split by cracks and gullies. A cast iron railway was planned, surrounding the base of the tower and radiating from it to the various landing sites. Iron trolleys carrying the blocks used this ingenious device which stood on iron legs, 5 to 6 ft proud of the reef and that enabled it to be laid almost level.

Work started for the second season on 26th May 1808 when *Sir Joseph Banks* dropped anchor off the reef. The upper portions of the wooden framework were quickly enclosed to form a covered refuge for the men incorporating a platform, 20 ft above the waves, for the blacksmith's forge and for the mixing of cement. The beacon would now serve two invaluable functions; the blacksmith was able to work unhindered by the waves which previously had doused his forge with untimely regularity, and more importantly, the shelter it afforded made the wearying journeys to and from Arbroath unnecessary and offered the opportunity to work for every possible moment on the tower.

Left: This view of the lantern from an NLB helicopter shows exactly how the previous photograph was taken. The ladder to the weather vane and anemometer leans against their support, although a head for heights was clearly needed if any work was to be done on the instruments. Bell Rock used to be a Met Office coastal reporting station. *(Author)*

Throughout June large numbers of men worked feverishly between the tides to excavate the footings of the light and to erect the railway from their workings to the various landing points. Within the circular base of the tower any irregularities of level caused by fissures were filled by 18 variously shaped blocks which were loaded on to row-boats and dropped over the site at high water as the railway was not yet complete. Once in position, a circular base 42 ft across with a perfectly smooth floor was ready to receive the first complete course of 123 stones. The first was set on 10th July and the remainder occupied the rest of that month. Every block

was cut following Smeaton's precedent on Eddystone – dovetailed to its neighbour and held in position using wedges, joggles and trenails until its security was finalised by the setting of the mortar or by the laying of another course on top of it. In this first course 508 cu ft of granite and 76 cu ft of sandstone were used, weighing together 104 tons.

Above: A striking study of Bell Rock lighthouse from the east. Notice how the green staining, caused by wave action, extends almost half way up the tower!
(Peter J. Clarke)

A second course of 136 stones, weighing 152 tons, was laid by 27th August, which brought the solid base 2½ ft from its foundations and clear of the reef, thus removing the time-consuming and monotonous pumping of the workings before the day's labours could be commenced. By 21st September the third course was complete, but heavy weather forced Stevenson to call a halt to the proceedings. Over 400 stones had been set upon the reef to give a pillar of granite which rose almost 5 ft from it and weighed 388 tons.

During the close season two more vessels were acquired or built in an effort to ensure the continual supply of stones into the workyard and the dressed blocks from there to the reef. Every stone, upon being finished, was laid out in the yard to check its fit with that of its neighbour. If satisfactory, the grooves and holes for the wedges and trenails could be cut.

The new season started badly, towards the end of April. By 31st May, six days after the first stones of the fourth course had been landed, 3 ins of snow lay on the workings. Fortunately, it did not last, and work on the wooden beacon could go ahead apace. A ropeway was slung between barrack and tower giving access before the tide had receded fully. By 25th June, the seventh course was in place, thanks to the efforts of over 50 men who worked solidly through both tides. At one point during the summer months the blocks were actually being laid at a faster rate than they could be delivered, so much so, that by 8th July it was noted that the high tide failed to cover the stump of the tower, now 13 ft above its base. This was the signal for much rejoicing and decorating of the masonry and attendant vessels with flags. A three-gun salute was fired, to the hearty cheers of all concerned.

By lodging in the three-tiered barrack, every single opportunity could be taken to press ahead with the work. During August the tower rose with tremendous rapidity, sometimes at the rate of over 50 blocks a day, until on 25th August all 1,400 tons of the solid base were finally complete. It had taken three seasons to get just this far, 31 ft above the rock and 17 ft above high spring tides. This was, Stevenson concluded, a fitting point at which to call a temporary halt to the upward progress, although 24 men stayed behind until November to complete the railway and strengthen the legs of the beacon.

By the April of 1810 work had started once more, mainly repairing storm damage to the railway and fixing a rigid timber walkway between barrack and tower. The first batch of stones arrived on the *Smeaton* on 18th May. Throughout June and July, frenzied activity from the men was the order of the day as Stevenson realised that it was within his grasp to complete the masonry by winter. Several times gales swept over the unfinished tower and halted their progress, but by 9th July the final stones were unloaded on to the reef ready to be hauled almost 100 ft up the side of the tower. On the 30th of that month the last stone was gently lowered into position to complete the 91st course, accompanied by a statement from Stevenson: "May the great Architect of the universe, under whose blessing this perilous work has prospered, preserve it as a guide to the mariner." The masons returned to Arbroath and were replaced by carpenters engaged to complete the internal fittings. Doors, window frames and ladders were positioned, and by 25th October the Argand burners were installed in the lantern. Finally, as the culmination of three days continuous labour, two 5 cwt fog bells were hung.

However, it was not until 1st February 1811 that the lighthouse entered service owing to difficulty in obtaining large sheets of red tinted glass for the lantern. This accomplished, a notice was inserted in various newspapers and journals to the effect that:

A lighthouse having been erected upon the Inch Cape or Bell Rock, situated at the entrance to the Firth of Forth and Tay, in north lat. 56°29´, and west long. 2°22´, the Commissioners of the Northern Lighthouses hereby give notice that the light will be from oil, with reflectors, placed at a height of about one hundred and eight feet above the medium level of the sea. The light will be exhibited on the night of Friday, the first day of February 1811, and each night thereafter, from the going away of daylight in the evening until the return of daylight in the morning. To distinguish this light from others on the coast, it is made to revolve horizontally, and to exhibit a bright light of the natural appearance and a red-coloured light alternately, both respectively attaining their greatest strength, or most luminous effect, in the space of every four minutes: during that period the bright light will, to a distant observer, appear like a star of the first magnitude, which, after attaining its full strength, is gradually eclipsed to total darkness, and is succeeded by the red-coloured light, which in like manner increases to full strength and again diminishes and disappears. The coloured light, however, being less powerful, may not be seen for a time after the bright light is first observed. During the continuance of foggy weather and showers of snow, a bell will be tolled by machinery, night and day, at intervals of half a minute.

With the appearance of the notice nobody was more relieved than Stevenson himself. His devotion to his task throughout the four years had finally been rewarded. Four years which had seen some of the most foul weather the North Sea could produce, allowing only furious bursts of work before the rising tides or menacing seas drove them off the reef. In the end his determination had won through. He had produced a masterpiece – a solid pillar of stone rising 115 ft out of the sea, 42 ft across its base, tapering smoothly to only 15 ft at the top. In all, 28,530 cu ft of stone weighing 2,076 tons had been shaped, dovetailed and cemented into a monolith of true beauty. It had cost £61,331. Six rooms were provided, the lower three for coals, water and oil, the upper trio serving as bedroom, kitchen and sitting room. Above these was the cast-iron octagonal lantern room with its burners, reflectors and highly polished glass.

On that day in February 1811 when the lantern machinery was set in motion, a reassuring multicoloured flash greeted mariners close to the Forfarshire coast. Sadly, to overshadow his triumph, Stevenson's efforts were questioned in public some years later and, while it is generally accepted today that the architect and engineer of the Bell Rock lighthouse was Robert Stevenson, this has not always been the case. It appears that during the latter half of the 19th century there was considerable dispute between the Stevenson family and the son of John Rennie, who together with Stevenson and the famous Thomas Telford, were jointly responsible for the execution of the undertaking although not, it now appears, to the same degree.

The misunderstanding first came to light in about 1848 when Sir John Rennie published a paper in which he claimed his father, who had since died in 1821, designed and built the Bell Rock lighthouse. Stevenson's sons promptly denied this claim in literature of the time, yet it again sprang to prominence when a book entitled *Lives of the Engineers* by Samuel Smiles appeared in 1874, even though when Stevenson died in 1850 the Commissioners of Northern Lighthouses honoured their great engineer by stating that to Stevenson "...is due the honour of conceiving and executing the great work of the Bell Rock lighthouse."

It was apparently from a letter of March 1814 written by John Rennie to a friend, claiming that he prepared the plans and visited the site on occasions, plus minutes of the Northern Lighthouse Board proclaiming Rennie as chief engineer, that Sir John based his claim. While it is true that Rennie prepared a plan for a lighthouse, it was in fact never adopted, the design of the tower is unquestionably Stevenson's. Also, it would appear from correspondence of Rennie's that, for various reasons, on the only three occasions he visited Arbroath to see the stones being cut, he never actually set foot in the tower while it was being constructed, despite travelling out to the reef and landing during two of the visits.

There exists much more documented evidence to suggest that Sir John's statement

Above: Huge rollers break against the Inchcape Rock from the east. The iron walkways are seen to good effect in this shot. *(Jim Bain)*

about his father, while not being a complete figment of his imagination, certainly seems to fall down under close investigation. Much of this lengthy documentation appears in an appendix to a book entitled *The World's Lighthouses Before 1820* by D. Alan Stevenson, himself a descendant of Robert. Even so, the evidence is presented in such a concise and unbiased manner that it leaves the reader in little doubt as to whose brain-child the Bell Rock lighthouse was. Most modern sources concur with the conclusions reached in this appendix and honour Robert Stevenson as the creator of this great work of civil engineering.

This is not the only time that the name of Stevenson occurs in this volume. Indeed, the Bell Rock heralds the start of an era of lighthouse construction around Scottish waters that is dominated by the Stevenson family. Many of Robert's descendants maintained the standard set by their father to build towers whose beauty and grace equalled, if not exceeded, that of the Bell Rock. Trinity House boasted that the name Douglass was connected with its finest works, while north of the border the Stevenson family reigned supreme, acclaimed by all who knew their work as the finest lighthouse engineers throughout the world.

The history of the Bell Rock lighthouse is a long one. Nearly 200 years have elapsed since the first stone was set upon the reef and it is now the oldest remaining rock tower in existence around the British Isles. There is no other rock station where the original tower still functions without being superseded by a more modern structure. For nearly 200 years it has resisted tempest and storm to remain as sound as the day it was built. Its durability or strength have never been called into question.

Obviously, with such an extended history its lonely vigil has not been without incident. On 30th July 1814, only three years after its operations commenced, it was visited by Sir Walter Scott while on a tour of Scottish lighthouses. After partaking of breakfast with the keepers he wrote the now famous verse in the visitors book:

Pharos loquitur
For in the bosom of the deep
O'er these wild shelves my watch I keep,
A ruddy gem of changeful light,
Bound on the dusky brow of Night:
The seaman bids my lustre hail,
And scorns to strike his timorous sail.

It was also in 1814 that Stevenson decided the outside of his lighthouse should be painted white, a decision prompted by the discolouration of its walls.

The lantern room of the Bell Rock has also witnessed many modifications – most intentional, but with the occasional unexpected remodelling! For instance, it is well known that the brilliance of a lighthouse often acts as a curious magnet for seabirds. On the night of 9th February 1832 a herring gull dashed against the windows of the lantern room with sufficient force to shatter the glass into thousands of fragments. The keepers, hearing the splintering glass, rushed to examine the cause of the damage and found a herring gull, with an estimated wingspan of 5 ft, on the floor of the lantern room. A sizeable herring was lodged in its gullet, along with a piece of plate glass.

In 1842, the entire contents of the lantern (lamps, reflectors and machinery) were removed and shipped to Newfoundland where they continued to give excellent service in Bonavista lighthouse for another thirty years. The replacement machinery in the Bell Rock was converted to use paraffin as an illuminant instead of colza oil in about 1877.

By 1890 the lighthouse could warn of fog using a tonite (an explosive made from gun cotton and barium nitrate) explosive fog signal which replaced the two fog bells. On 5th April that year an explosive charge detonated prematurely causing considerable damage to the lantern windows and the optics themselves. It took until 13th April before repairs could be effected and the light returned to service – the first time since it was built the lighthouse had failed to give its nightly warning.

There were further modifications in 1902 when the entire lantern was removed and all the equipment replaced by what has been described as, "one of the finest lenticular apparatuses then made." The lenses had equiangular glass prisms with a focal distance of 1330mm emitting a red and white flash every 60 seconds. During these alteration the two obsolete 5cwt fog bells were removed, one of which was donated to Arbroath Museum were it can still be seen today in the Signal Tower Museum.

The major modifications of 1964 improved the optics and upgraded the living accommodation for the keepers, this having changed very little since the lighthouse was built. At this time Chicken Rock lighthouse off the southern tip of the Isle of Man was being automated and its original lenses of 1875 would become redundant. It was decided to install these in the Bell Rock. The paraffin vapour burners were removed and replaced with a 3,500 watt electric light bulb which produced 1,900,000 candlepower in a single white flash every

Left: The *Summer Rose* from Arbroath was the Bell Rock supply boat, although on this day there is not much to supply as the lighthouse is being converted to automatic operation. The solar panels are in position on the gallery, but there is no lens in the lantern. Two temporary lanterns have been stacked on top of one another on the roof of the dome. The flags decorating the tower are to celebrate its 175th anniversary. *(Jim Bain)*

three seconds that was visible for 28 miles. The electricity was produced by generators installed inside the tower. Fresh water and fuel oil capacities were increased by additional tanks, some of which are on the lantern gallery, the tonite fog warning system was replaced by three supertyfon compressed air generators, while the rooms within the tower had their functions changed around. From top to bottom they became; lightroom, control room, living room, bedroom, store room, upper engine room, lower engine room, access shaft and entrance. The old lightroom had all the optics installed on an upper level, while all the complex electrical control and monitoring gear was installed below this.

A similar incident to the 1890 tonite explosion took place many years later during World War II, although its cause was rather more deliberate. On April 1940 an enemy shell was dropped onto the reef, exploding about 10 yards from the base of the tower, but near enough to destroy several lantern panes. This, together with cursory machine-gun attacks at intervals during the hostilities, resulted in 9 bullet holes through the dome, 14 lantern panes broken, 4 damaged lens prisms, 6 red shades smashed, plus damage to the balcony and balcony rail. Fortunately, there were no fatalities.

During the First World War a rather unfortunate chain of events combined to produce what were nearly fatal consequences. Normal practice during wartime was for the lighthouse to be in darkness, except when specifically requested to be otherwise. On 27th October 1915, the Captain of the *Argyll*, an armoured cruiser of 10,850 tons, sent a message to the Admiral Commanding the Coast of Scotland at Rosyth requesting the Bell Rock to be lit on the night of October 27th/28th. All messages to the lighthouse had to be delivered by boat as no radio was installed, but unfortunately the message never reached the keepers owing to heavy seas. The *Argyll* foundered on the reef but the crew of 655 were saved.

The crew of an unauthorised helicopter delivering newspapers to the keepers in December 1955 were not so lucky. The rotor arms caught the top of the tower plummeting the craft into the sea. The crew died as a result of their misjudgement while the damaged lantern made the lighthouse unserviceable for a week.

The lighthouse remained unserviceable for a considerably longer period after a disastrous fire on 3rd September 1987. A spillage of diesel fuel oil, the fumes from which ignited, caused a severe fire in the kitchen and required the keepers to evacuate the tower completely. On their way down through the tower they wisely closed hatches behind them and threw out any inflammable materials left by contractors working in the tower, thus minimising the effects of the fire. The three keepers were air lifted from the entrance balcony into a Search and Rescue helicopter from RAF Leuchars without injury.

The greatest damage occurred in the kitchen where the mainly wooden fittings burned fiercely. The heat was sufficiently intense to crack the internal granite walls (some of the cracks extending through to the outside), blow out the windows and melt the aluminium ladder to the lightroom. As the fire spread upwards into the lightroom it destroyed the lantern glazing, radio equipment and batteries, as well as being hot enough to blister the paintwork on the outside of the lantern dome. A temporary light and two 'racons' mounted above the dome survived. Immediately below the kitchen, the bedroom suffered smoke damage.

If ever such a disaster could be described as 'fortunate' this might be an appropriate description of the Bell Rock fire. The lighthouse was in the course of conversion to automatic operation and was at the interim stage between the removal of the 'old' equipment and the installation of the automatic machinery. The lost equipment was therefore of little consequence, apart from the fact that the programme of automation was somewhat delayed.

A temporary kitchen was installed in the room below the bedroom while refurbishment commenced. The tower was made weatherproof and remanned on 6th November 1987. Contractors filled the internal cracks by pressure injection of concrete and reglazed the lantern, enabling the automation programme to recommence early in March 1988. An interesting angle to the conversion progress was the installation of solar panels around the lantern gallery to provide power for rotating the optics and powering the 'racon' and radio monitor. An acetylene gas burner provides the light.

The Bell Rock has never had a 'helideck' as such but today it does have a huge netting shroud that looks a bit like the Trinity House style of helideck supports, draped over the outside of its lantern. Its purpose is to prevent migrating sea birds perching on the solar panels, fouling them with droppings and reducing their efficiency. The final keepers of the Bell Rock lighthouse were withdrawn on 26th October 1988, since when it has been monitored from the NLB Headquarters in Edinburgh.

It would be fair to say that the building of the Bell Rock lighthouse was a landmark in this field of civil engineering. Its success was due, not only to the genius of its designer, but also to the enthusiasm and dedication of the men he had in his charge. By insisting on minimum periods for working on the rock, and paying particular attention to the welfare of these men, Stevenson was able to inspire and lead them to the rapid completion of their project, despite the seemingly overwhelming factors against them..

Its construction was watched with awe by sceptics on the Scottish coast, just as today's generations marvel at its slim profile breaking the horizon 11 miles off Arbroath. While they watched the slender stalk of masonry grow they were probably unaware that this was going to be the last major rock tower built with the aid of sailing vessels, for the Age of Steam was soon to revolutionise civil engineering.

Stevenson saw the Bell Rock operate for less than 40 years before he died. If he could be remembered for just one achievement then it would surely be this lighthouse. Its ageless defiance of the waves is matched only by the dedication of its builder.

8 SKERRYVORE
the noblest of all

IT WAS with some affection that I recount the history of the lighthouse known as Skerryvore. Some years ago, in my youth, Skerryvore was the first rock light I set eyes upon, not at close quarters, but from a beach on the corner of the Hebridean isle of Tiree. It was 11 miles away, but even at that distance, white spray breaking against the base of a tapering silhouette could clearly be seen. That night, the wheeling rays from the lantern split the night sky, a giant spoke of light sweeping rapidly over that windy isle. From then onwards, with the fleeting glimpse from the shore planted firmly in my memory, I became fascinated by Skerryvore and all such similar structures, marvelling at the courage and dedication of the men who built them, while imagining their titanic struggle with the elements to accomplish their end.

My baptism into the world of pharology could not have started with a finer example. I still believe there is not a tower in the world which can surpass Skerryvore in its beauty of design, a beauty which has justifiably earnt this structure the accolade "the noblest of all extant deep sea lights."

It stands on the high point of a shoal of rocks extending for no less than 8 miles, to the south-west of Tiree. They are volcanic remnants, thrown up millions of years ago, since when they have been battered mercilessly by the crashing Atlantic into a lengthy barrier of islets and skerries, now the home for colonies of sea birds and seals. In many places these dark rocks break the surface as sizeable humps. Indeed, some are never covered by normal tides, but usually the more evil of these teeth are only shown at low water. Some are never seen at all, their position betrayed only by the broken and foaming water amongst the larger rocks. There is no other reef around our coastline which extends for such a length or is fraught with the dangers which beset this particular barrier to navigation.

For over 8 miles the shoal effectively prevents an easy passage through the already difficult waters of the Hebridean sea. Vessels out of the thriving port of Oban making for open waters either take a route well to the north of Tiree and Coll, or through the confined passage between them, actions imperative to avoid this unkindly line of rocks. There were many sea captains who would not attempt even these diversions during the hours of darkness lest an error of navigation or heavy seas swept them off course and into the jaws of this waiting demon.

Between 1790 and 1844 over 30 ships were known to have been lost in this area. This figure does not include those of which no trace was found and were written off as being 'lost at sea'. There are some accounts, which may well be apocryphal, that rents for property and land in the south west corner of Tiree were higher than on the rest of the island because of the amount of flotsam that could be scavenged from the beaches after a shipwreck on Skerryvore reef. The Commissioners of Northern Lighthouses had long been aware of this lurking menace; in 1804 Robert Stevenson, in his capacity as chief engineer to the board, landed on the reef and confirmed the need for some kind of beacon to be placed there. Ten years later, in 1814, and having finished his great work on the Bell Rock, he was to return, this time with Sir Walter Scott and a party of Commissioners. He made detailed measurements to assess the feasibility of such an undertaking. The writings of Scott after the event give a vivid impression of the difficulties experienced that day.

Quiet perseverance on the part of Mr S, and great kicking, bouncing, and squabbling upon that of the yacht, who seems to like the idea of Skerry Vhor as little as the Commissioners. At length, by dint of exertion, come in sight of this long ridge of rocks (chiefly under water), on which the tide breaks in a most tremendous style. There appear a few low broad rocks at one end of the reef, which is about a mile in length. These are never entirely under water, though the surf dashes over them. To go through all the forms, Hamilton, Duff and I resolve to land upon these rocks in the company of Mr Stevenson. Pull through a very heavy swell with great difficulty, and

Opposite: The magnificent proportions of Skerryvore lighthouse (*Author*)

Right: The true extent of the reef on which Skerryvore stands is a little more obvious in this photograph, although there is still a considerable amount we can't see. The "white fields of foam" that Alan Stevenson remarked upon during its construction are clearly visible. *(Scotsman Publications)*

approach a tremendous surf dashing over black pointed rocks. Our rowers, however, get the boat into a quiet creek between two rocks, when we contrive to land well wetted. I saw nothing remarkable in my way except several seals, which we might have shot, but in the doubtful circumstances of landing we did not care to bring guns. We took possession of the rock in the name of the Commissioners, and generously bestowed our own great names on its crags and creeks. The rock was carefully measured by Mr S. It will be a most desolate position for a lighthouse, the Bell Rock and Eddystone a joke to it, for the nearest land is the wild island of Tyree, at 14 miles' distance. So much for the Skerry Vhor.

In the same year that Scott made his visit, an Act of Parliament was passed providing for the erection of a lighthouse on Skerryvore. Attendant difficulties connected with the reef, particularly its isolation, delayed matters. It was only 11 miles from Tiree, yet over 50 from the nearest mainland. Eventually, in 1834, Stevenson returned again, this time with his 27-year-old son, Alan, to whom would be entrusted the Herculean task of lighting this reef. He made a detailed survey of over 130 of the main rocks before leaving to decide on his plans. It was soon apparent that Stevenson had little real choice as to the site or the material for his lighthouse. The largest hump of the reef was about 280 sq ft in area at low water, and was never submerged by normal tides. Wave pressure measurements in excess of 6,000 lb per sq ft immediately confirmed in Stevenson's mind the idea of a masonry tower as "no pecuniary consideration could in my opinion

Left: A Force 10 at Skerryvore – huge waves leap into the air as they hit the reef. *(Northern Lighthouse Board)*

have justified the adoption of an iron lighthouse for Skerryvore."

So the rock called Skerryvore (Gaelic, *sgeir* – a rock, *mhor* – big) was chosen, and Alan Stevenson wisely decided to set a masonry tower upon it. Only in this manner could a structure with enough mass to endure the force and weight of water thrown against it survive. Its final weight of 4,308 tons acted almost vertically downwards and was a key factor in Stevenson's reckoning. His plans, when complete, were for the most daring structure yet seen; a tower 138 ft high whose base spanned 42 ft, narrowing to just 16 ft at the lantern gallery. The lowest 26 ft were entirely solid, the cubic capacity of this base alone being double the entire structure on the Eddystone, while the complete tower would be 4½ times greater than Smeaton's work and twice that on the Bell Rock. It was the tallest,

SKERRYVORE LIGHTHOUSE.

Left: An early engraving of Alan Stevenson's magnificent Skerryvore – although the artist has given the stonework something of a dry stone wall quality.

heaviest and most graceful tower yet attempted, and it was to be built by a man of only 27 who had no experience of wave-swept towers to guide him. However, he was a Stevenson, and if anyone was going to complete such a gargantuan task it would be a member of that family.

By 1837 the first tentative steps were being taken. A pier, workyard and harbour were started at Hynish on Tiree where locally quarried granite was used, together with stone from Mull, generously donated by the Duke of Argyll who gave Stevenson a free hand to take stone from any of his estates. During the summer of 1838 £15,000 was used in wages for the construction of a 150 ton steamer at Leith for ferrying men and stone to the reef, 12 miles from Hynish, and also for the construction of a wooden barrack building to lodge the masons when at work on the rock. This was similar to the barrack that was first seen on the Bell Rock, a six-legged, open framed pyramidal structure, on top of which sat the wooden shelter containing kitchen, store rooms, and sleeping quarters for up to 40 men. The design had served well in the North Sea and would be ideal on Skerryvore where the waters are seldom tranquil enough to allow the men to lodge permanently in a floating barrack. It was built in a workshop in Gourock during 1838 before being dismantled and despatched to the reef.

In the same year, Hynish was turned into a hive of activity – a collecting point for men, boats, tools, cranes, winches, forges, masons and blacksmiths, all converging on the tiny island settlement where new buildings were rising rapidly. Thirty masons and about four blacksmiths had made the long journey from Aberdeen, from where Stevenson could be sure that if nothing else they would be expert in cutting granite. On 23rd June, Stevenson left Tobermory on Mull to start his great work.

While awaiting delivery of the steam tender, he made use of the *Pharos*, the sailing vessel which had given invaluable service to his father while constructing the light on the Bell Rock. On 28th June, with the *Pharos* anchored off the reef, Alan Stevenson leapt from a pitching row-boat on to Skerryvore, where he had ample time to ponder on his fate should the boat be unable to return. Soon, however, it did return with a gang of workmen who immediately busied themselves with the marking out of positions for the tower and wooden barrack. The hard black volcanic rock of the reef was extremely difficult to move over. Centuries of wave action had polished the surface to a glassy smoothness which was generally running with water. The main rock was deeply incised by gullies which, although they provided good shelter for small boats, restricted considerably the area on which Stevenson could plant his tower. One of these gullies terminated in a vertical spout of rock, up which water was driven at regular intervals. This caused Stevenson to comment that,

The effect of the jet d'eau was at times extremely beautiful, the water being so broken as to form a snow-white and opaque pillar, surrounded by a fine vapour in which, during sunshine, beautiful rainbows were observed.

Although doubtless beautiful, the constant spray made the smooth rock surface treacherous, especially when leaping from boats with legs and arms searching desperately for grip. The foreman of the masons likened the experience to climbing up the neck of a bottle.

On 30th June Stevenson left with the *Pharos* and a number of smaller vessels for Gourock to collect the wooden barrack. After a storm-tossed return journey he landed again on Skerryvore at midday on 7th August, when four carpenters, 16 stone masons and a blacksmith began unloading and securing their most valuable cargo. After a night on the *Pharos* work resumed the following day, but was interrupted that night by one of the many storms which Stevenson was to become so familiar with over the next few years. There was nothing for it except to return to Tiree and wait. Six frustrating days later the weather allowed them to return to the reef and they recommenced their work.

The massive legs of the barrack were set into holes blasted out of the rock. It was tedious and tiring work, occupying 16 hours of every day from 4 am to 8 pm. At 8 am boats would

arrive with pitchers of tea, biscuits and canteens of beef which were consumed in a half-hour break. Dinner was served at 3 pm with the addition of vegetables and beer to replace the tea. Work would continue until 8 or 9 pm when it was so dark they could hardly see to scramble into the boats. Once on board a further meal, supper, was offered, although many of the men,

*Right: An interesting shot of Skerryvore, this picture was produced from a single frame of cine film from a commercial by Philips to advertise their bulbs.
(Philips Electrical)*

unused to the peculiar motions of the *Pharos*, did not eat this. After such gruelling work on the rock few of the men could adapt to sleeping in the rolling vessel, and many preferred the more stable, if damper, rock surface for their rests.

Stevenson called a halt to the proceeding on 11th September 1838 when the legs of the barrack had been securely fixed into the rock and to one another. They had worked only 165 hours in 35 days. It was slow progress which meant the habitable turret would have to be left

until the following spring. Before the final departure of the season a wooden chest packed with a water cask and biscuits was bound to the top of the pyramid in case any unfortunate sailors were wrecked on Skerryvore during that winter. In the execution of this humanitarian gesture Stevenson was able to gaze from the apex of the legs.

I now for the first time got a bird's eye view of the various shoals which the stormy state of the sea so well disclosed. And my elevation above the rock itself decreased the apparent elevation of that rugged ledge so much that it seemed to me as if each successive wave must sweep right over its surface and carry us all before it into the wide Atlantic.

Stevenson spent the winter in Edinburgh. He knew from the progress already made that he would be returning to Skerryvore every spring for many years to come. Yet he was optimistic, and believed that, "…the successful termination of our first season's labours might be taken as an omen of future success." He was wrong. On the night of 3rd November an Atlantic storm of seismic fury beat down on Skerryvore rock. It was probably a single wave which swept the legs of the barrack into oblivion. Not only that, a grindstone for sharpening the tools was picked up and dumped at the bottom of a 3 ft deep hole, the iron anvil was moved 13 yards from where it was left, a half hundredweight block lying in an excavation was hurled into the air to rest on the highest point of the rock, and the iron stanchions which anchored the legs into the reef were twisted like corkscrews. Clearly, little could have survived such an assault. The fruits of a whole season's labour were destroyed in a single night.

Over a week later, on 12th November, Stevenson received a letter from his clerk of

Above: On 16th March 1954 a disastrous fire swept through Skerryvore lighthouse putting it out of action for over 5 years. We can see the smoke-blackened and shattered lantern panes and blackened masonry around the windows and doors a few days after the event. In the background is the NLB tender *Hesperus.* (*Glasgow Evening Citizen*)

Far left: Dramatic evidence of the inferno that swept through Skerryvore on 16th March 1954. The plate glass lantern panes are blackened or shattered by the intense heat that funnelled up through the tower. (*Glasgow Evening Citizen*)

Left: New railings around the gallery and a completely new lantern and lens were required to return Skerryvore to full operation after the fire. Notice how the four rows of glass panes in the lantern before the fire have been reduced to just two afterwards. (*Author*)

works, a store keeper on Tiree. "Dear Sir," it began, "I am extremely sorry to inform you that the barrack erected on Skerryvore rock has totally disappeared." Such news came as a shattering blow to his confidence, but on the same day as he received the letter a boat was hired in Glasgow to take him to the reef to inspect the damage personally. What he saw in no way altered his belief that what he was doing was the only way to conquer Skerryvore. He would return to build an identical, but stronger, structure, in order to continue his battle with the sea. In the face of defeat Stevenson stood firm. For the engineer, the spring of 1839 must have been an eternity in arriving.

When Alan Stevenson stepped onto Skerryvore during April 1839 he was faced with two overriding problems, both of which had to be conquered if he was to complete the lighthouse on time. Firstly, there was a new barrack to be built, and secondly, while this was being raised work must also be started on the foundation for the tower. It is a tribute to Stevenson and his men that both were accomplished during this year. The men worked feverishly to complete the barrack as soon as possible – any kind of stable shelter was to be desired in preference to the giddy motions of the attendant vessels. When eventually complete, on 3rd September, the barrack stood 60 ft above the reef and was entered by climbing ladders attached to the legs. It contained three separate areas, one on top of the other. The lowest served as kitchen, the middle storey was divided into two cabins for Stevenson and his master of works, while the 30 to 40 men had to billet in the upper floor, where conditions were, to say the least, cramped.

Before stones had been laid, this was the only structure on the reef. It was occupied several months before all fixtures and fittings were complete, especially those to make it wind and water tight. During these early days the inhabitants suffered frequent flooding, especially during northerly gales. They also had difficulty in keeping warm. When wind or waves prevented labour on the reef, days on end could be spent in the confines of the shelter waiting anxiously for favourable conditions which would permit their descent on to the rock.

During such conditions communications with Tiree were subject to much disruption. All mails, materials and stores had to be landed from boats which could not approach during inclement weather. On one occasion during 1840, 14 days elapsed before the reef could be approached by a steamer. The men were growing noticeably short of provisions as they gazed expectantly at the distant horizon. "We saw nothing but white fields of foam," Stevenson describes, "as far as the eye could reach. For several days the sea rose so high as to prevent our attempting to go down to the rock."

The foundations of the tower occupied the entire season of 1839, from 6th May to 30th September. A giant circle, 42 ft in diameter, was marked out and all material within this line removed. This gave the largest possible base without a change in level. No fewer than 296 charges of dynamite were used to remove 2,000 tons of rock, a rock described by Stevenson as, "a hard and tough gneiss, requiring the expenditure of about four times as much labour and steel for boring as are generally consumed in boring the Aberdeenshire granite." Progress was limited by the time taken to resharpen their rapidly blunted tools, while the gunpowder had to be used with extreme caution as it caused the gneiss to splinter like glass in all directions. There must have been many anxious moments on a site where shelter from such flying debris was scarce, although it is to Stevenson's credit that no fatalities occurred during this or subsequent work on the lighthouse. The excavations employed 20 men for 217 days in this, and the following season, about which Stevenson commented:

If it be remembered that we were at the mercy of wind and waves of the wide Atlantic and were every day in the expectation of a sudden call to leave the rock and betake ourselves to the vessel, and on several occasions had our cranes and other tools swept into the sea, the slowness of our progress will excite less surprise.

By the end of the season the base was almost ready to receive blocks, an operation which would now have to wait until the following year. Stevenson had spent two seasons on Skerryvore and had not yet landed a block, only wood and metal stood on the reef. During the close season much activity was in progress away from the site. Between April 1839 and June 1840 quarries on Mull echoed to the sound of repeated blasting that yielded over 4,300 blocks from the island. These were roughly shaped, despatched to Hynish, where gangs of masons hammered and chiselled them to a shape determined by a template so that little extra work would be required when on the reef. The stones varied from ¾ ton to 2½ tons in weight, and could occupy anything from 85 to 320 hours of a man's time in shaping them to the degree of precision necessary.

Hynish, too, was bustling with activity. Not only were the blocks being shaped and fitted, but a harbour and pier were under construction. More accommodation was built to house the swelling numbers before construction of the lighthouse could begin in earnest.

The morning of 30th April 1840 was the start of the year's operations. The barrack, much to the relief of the engineer, had survived the winter's gales and was ready for occupation once more. By 20th June the first stones were arriving at the reef on board the appropriately named *Skerryvore* which was duly bedecked with flags for the occasion. The first stone was laid on 4th July 1840 by John, the seventh Duke of Argyll, from whose quarries the stone was being hewn. His visit on that day made an impression on his mind which could not be erased by the passage of time.

That sight is as fresh in my memory after an interval of 57 years as if I had seen it yesterday. The natural surfaces of the rock were irregular in the highest degree. Worn, broken and battered by the unnumbered ages of the most tremendous surf, and by the splitting of the rock along lines of natural fissures, there did not seem to be one square foot of rock which was even tolerably level.

Yet in the midst of this torn and fissured surface we suddenly came on a magnificent circular floor, 42 ft in diameter, as level as water, and as smooth as a billiard table. Its containing walls of living rock rose around

every portion of its immense circumference to varying heights, showing the various depths of cutting that had been needed to reach a perfectly solid level. On this floor the whole weight of the tower was intended to rest. And it was on this weight that the stability of the enormous structure was entirely to depend.

At last, a start had been made. The men, inspired by Stevenson, were eager to press on and gain every advantage while the weather held. Landing the blocks was by no means easy. The bigger vessels brought their cargo from Hynish and anchored a safe distance from the reef where the stones were unloaded into smaller lighters, capable of negotiating the narrow creeks where cranes were waiting to relieve them of their cargo. The men soon became expert at handling boats and cranes, although in such circumstances they could never feel

Above: A truly stunning shot of Skerryvore lighthouse by renowned lighthouse photographer Jean Guichard. The lighthouse is empty of human life, but the dangers of which it warns are only too apparent. *(Jean Guichard)*

totally complacent. A freak wave which could pitch men or stones into the turbulent creeks of Skerryvore was always to be expected. Indeed, many of the helmsmen of the lighters were roped into position in case such an eventuality arose.

The precision obtained at Hynish allowed good progress, sometimes at the rate of 95 blocks a day. When winter weather forced an abandonment of the reef, 800 tons of granite stood on Skerryvore, a solid pillar of 10,780 cu ft and rising 8 ft 2 ins from it.

As many as 80 craftsmen toiled through the winter on Tiree, dressing and numbering blocks ready for despatch. Stevenson landed on 13th May 1841 to inspect the equipment and barrack for damage. Thankfully, he found little, and after a short delay caused by rough seas and illness of a mason, the full complement of men was landed on 22nd May. The 8th July marked another milestone in the construction when the solid base was finally complete – 26 ft of immovable masonry upon which a balance crane, capable of lifting 2 tons at a time, was placed for setting blocks above this level. By 17th August a further 30,300 cu ft had been added – double the amount that existed in the entire Eddystone light.

The barrack was surviving well enough for Stevenson, during periods when work was impossible, to reflect upon life and conditions in his timber prison, and particularly to consider the tedium endured by himself and his men.

I spent many a weary day and night – at those times when the sea prevented anyone going down to the rock – anxiously looking for supplies from the shore, and earnestly looking for a change of weather favourable for prosecuting the works. For miles around nothing could be seen but white foaming breakers, and nothing heard but howling winds and lashing waves. At such seasons much of our time was spent in bed, for there alone we had effectual shelter form the winds and spray, which searched every cranny in the walls of the barrack.

Even in the comparative comfort of their beds, sleep was by no means certain.

Our slumbers, too, were at times fearfully interrupted by the sudden pouring of the sea over the roof, the rocking of the house on its pillars, and the spurting of the water through the seams of the doors and windows – symptoms which, to one suddenly aroused from sound sleep, recalled the appalling fate of the former barrack, which had been engulfed in the foam not 20 yards from our dwelling, and for a moment seem to summon us to a similar fate.

Right: The NLB helicopter is dwarfed by the proportions of Skerryvore as it approaches with an underslung load of equipment – probably during its conversion to automatic operation.
(Eddie Dishon)

During these nocturnal tempests no one slept through fear of being caught unawares by another collapse of the barrack and some of the men preferred to spend these nights in the trunk of the lighthouse, sleepless and shivering in the cold and wet. During one season a storm broke and raged for seven weeks almost incessantly, making it impossible for the supply boat to put out from its harbour. During this period further discomforts were heaped upon the marooned men. Food supplies dwindled to a very low level, fuel was exhausted and worst of all, the supply of tobacco ran out.

The dawn of 17th April 1842 saw Stevenson making his preliminary inspection of the season and finding large boulders in the uncovered room at the top of the tower, 60 ft above the level of the highest tides. On 19th May the men returned, their numbers swelled in an effort to speed the progress upwards. It was indeed a remarkable progress. Work was started on the 37th course, and by 25th July the final stone of the 97th and final course was set in position. Triumph for Stevenson had eventually arrived, an appropriate moment for the engineer to pay fitting tribute to the courage, dedication and accuracy of his men. The tower was 138 ft from base to parapet, yet nowhere was the diameter of each course more than one sixth of an inch out of true, and the height only one inch greater than the architect's previous calculations. Over 58,000 cu ft of granite had been fashioned into a most graceful structure whose walls enclosed nine rooms and tapered from 9½ ft to 2 ft in thickness. On 10th August the lantern sections started arriving, and the year's work ended with these being hauled into position and the glazing completed.

In January 1843 Alan Stevenson became chief engineer to the Northern Lighthouse Board so many of the final details of the Skerryvore were placed in the capable hands of his brother Thomas, who arrived on the reef on 29th March 1843. To his pleasure, the whole tower had remained watertight during the winter, so he busied himself with attending to the interior fitments and repointing the external joints, by balancing precariously in cradles lowered down the outside walls from the lantern gallery.

The arrangement of rooms was very much as standard; entrance was gained through a gun metal door 26 ft above the rock, above which were found lower and upper engine rooms, workshop, living room, store room, lower and upper bedrooms, fog signal room, and service/control room. All were 12 ft in diameter and connected by oak ladders. Finally, the lantern apparatus was installed, Skerryvore's 'crown of glittering glass', an elaborate arrangement of mirrors and prisms.

The illuminating apparatus is revolving, fixed prisms below, eight panels of lenses revolving, and eight smaller panels also revolving above, to concentrate the upper rays; these are thrown on eight plane mirrors, which deflect them to the horizon parallel to the rest of the beam. The light is, therefore, a fixed light of low power, varied by strong revolving flashes. The lamp has four wicks, and is worked by pumps which ring a small bell when in action. The lamp machinery is wound up every hour and a half, and the keepers wind the revolving machinery at the same time, though it will go for three hours. The oil is hoisted up to the top of the tower by a movable crane, the water pumped up by a force pump. All the rooms have bell-signals, worked by blowing tubes, so that the keepers can call each other without leaving the lantern.

Even in these final stages the Atlantic refused to give in gracefully but, as Stevenson mentions in his narrative, marooned the men for seven long weeks in this storm-tossed bastion, all but exhausting their supplies of food and tobacco, not to mention wearing their clothes to rags.

However, in 1844, the Northern Lighthouse Board issued its customary notice to the effect that, "...a lighthouse has been erected upon the Skerryvore Rock, which lies off the Island of Tiree in The County of Argyll, the Light of which will be exhibited on the night of 1st February 1844 and every night thereafter from sunset to sunrise..." When the final accounts were produced they revealed the total cost for the lighthouse, the pier at Mull, dwelling houses, harbour and signal tower on Tiree to be £86,977-17s-7d.

The fame of Skerryvore and its lighthouse spread rapidly. Robert Louis Stevenson, himself a relative of the builder, was the first to acclaim this structure as, "the noblest of all extant deep sea lights" and went on to name his house in Bournemouth after it. *The Times* recognised Skerryvore as one of the four most prestigious towers in the world, "...the most perfect specimens of modern architecture. Tall and graceful as the minarets of an Eastern mosque, they possess far more solidity and beauty in construction, while, in addition, their form is as appropriate to the purpose to which they were designed as anything ever built by the Greeks."

With such publicity it is no surprise to find that Skerryvore was visited on many occasions by dignitaries who wished to view this feat of engineering at close quarters. William Chambers, Lord Provost of Edinburgh, stepped on to the reef in 1866 and paid tribute to its, "remarkable dimensions and matchless beauty of design," and recorded some interesting details about communication with Tiree.

As the weather had partially cleared we had a pretty extensive view over the wastes of waters from the balcony. The only visible land was that of Tiree at Hynish, with its signal tower. I was interested in knowing the method of intercourse by signals. Every morning, between nine and ten o'clock, a ball is hoisted at the lighthouse to signify that all is well at the Skerryvore. Should this signal fail to be given a ball is raised at Hynish to enquire if anything is wrong. Should no reply be made by the hoisting of the ball, the schooner, hurried from its wet dock, is put to sea and steers for the lighthouse. Three men are constantly on the rock, where each remains six weeks, and then has a fortnight on shore: the shift, which is made at low water of spring-tides, occurs for each in succession, and is managed without difficulty by means of the fourth spare keeper at Hynish, who takes his regular turn of duty. According to these arrangements, the keepers of the Skerryvore are about nine months on the rocks, and about three months with their families every year. But this regularity may be deranged by the weather. One of the keepers told me that last winter he was confined to the rock for 13 weeks, in consequence of the troubled state of the sea preventing personal communication with the shore. I enquired how high the waves washed up the side of the tower during the most severe storms, and was told that they sometimes rose as high as the first window, or about 60 ft above the level of the rocks; yet, that even in these frightful tumults of wind and waves the building never shook, and no apprehension of danger was entertained.

For over a century, the light from Skerryvore radiated across the Hebridean Sea until the events of a March night in 1954 ended its unbroken run. A disastrous fire occurred on the evening of 16th March which started on the seventh floor and spread rapidly downwards. The three keepers had no time to raise the alarm before they were driven on to the rock itself. Fortunately the relief vessel was due the following day and one evening's discomfort was all the three men had to endure, kept company, it is said, by a guillemot – supposedly the spirit of a former keeper – which sat around the trap-door openings. The intense heat had also caused fog signals to detonate inside the tower resulting in lengthy cracks in the masonry, one being over 100 ft long. By 24th March a light vessel was on station 4 miles from the reef, and in August the first of a series of temporary lights was installed in the tower itself while reconstruction work took place. Most significant was the installation of three generators to power the new electric light source, and the replacement of the damaged cast-iron lantern gallery with a modern anodised aluminium structure. The new light, exhibiting one flash every ten seconds, was first seen on the night of 6th August 1959 and has given reliable service until automation came to Skerryvore, over 30 years later.

In common with many other rock stations, the relief of the keepers at Skerryvore before automation, was effected by helicopter. Thankfully, there was room to construct a helipad adjacent to the base of the tower on the reef in 1972, thus eliminating any possible need for a helideck to be erected above the lantern, a sight now common on the rock lights of Cornwall but one which would completely desecrate the shapely form of Skerryvore.

The process of automating Skerryvore was a particularly lengthy one that was beset with a rash of varied problems. It all began in the late spring of 1991 when buoys were laid to the east and west of the tower, the keepers withdrawn, and the light extinguished. The first process was the not inconsiderable task of removing all the blue asbestos that had been applied to the inner walls of the tower and subsequently covered by wall boarding after the fire of 1954. A specialist firm of asbestos removers was employed to remove the material; part of the contract involved erecting a huge platform on the reef, supported by scaffolding, that was high enough to reach the entrance door. Onto this extensive promenade were located an engine room containing two diesel generators, a diesel storage area, a mess room, a general store, a radio/watchroom and a decontamination unit with water storage. A base camp was established on Tiree and daily helicopter trips were made to the island with the waste asbestos. How ironic that in its later life Skerryvore lighthouse again had a 'barrack-like' structure erected next to it on the reef which housed a workforce who were again subject to daily visits from a base camp on Tiree – just as it had been during its construction over 150 years earlier. Only the mode of transport and purpose of their work differed.

The lighthouse was entirely emptied of its contents, save for a large stainless steel tank that would later serve as a fuel tank. During late 1991 and early 1992 new living accommodation was re-established together with a temporary light with a 15-mile range. A permanent light produced by a 250W metal halide bulb with a nominal range of 23 miles was in operation again on 2nd October 1992, followed later by a automatic fog detector and signal. It was expected to finish installation of the new electrics by the end of 1992, followed by redecoration of the interior, although it wasn't until 31st March 1994 that Skerryvore was declared to be

Left: To the bottom of the frame we can see one of the jetties that were used for boat reliefs of Skerryvore. Thankfully there was enough room on the reef for a helipad which meant there was no need for a helideck above the lantern. *(Ian Webster)*

working automatically, and even then not all final work was complete inside the tower. The lighthouse is monitored by radio link to Ardnamurchan lighthouse and then via a landline to NLB Headquarters in Edinburgh. Almost 150 years to the month since Skerryvore first showed its guiding light a new phase in its history commenced.

As I remarked at the beginning of this chapter, I have yet to set eyes on a finer example of lighthouse engineering than the superlative Skerryvore. True, there are sites around our coasts that boast of lighthouses with equally dramatic histories, yet none of these can match the timeless grace of Skerryvore. Even the Institute of Civil Engineers were moved to describe Skerryvore as "...the finest combination of mass with elegance to be met within architectural or engineering structures." This was the pinnacle of Alan Stevenson's career – a reflection of the true genius of the engineer who not only produced a functional lighthouse, but managed at the same time to construct a thing of beauty.

Stevenson retired prematurely from his post as engineer to the Lighthouse Board due to ill health in 1853. He died in 1865. There can be no finer memorial to this great man than that which stands on a rock in the Atlantic 12 miles from Tiree.

9
BISHOP
ROCK
*the
Blue Riband
light*

THE FIRST landfall encountered in the western approaches to the English Channel are the Scilly Islands, a cluster of wind-swept, wave-lashed islets 27 miles to the south-west of Land's End. Their notoriety amongst mariners is legend, for there is probably no other region within the British Isles which has claimed more vessels within such a comparatively small area.

Study of a map reveals over 150 islands in the group, of which only five are inhabited. The majority are of no great size and often appear as sea-weathered, partly-submerged ledges and reefs of the hardest granite. Ships without number have been lost amongst these terrible rocks, driven to their fate by the untamed fury of an Atlantic gale as they searched vainly for a safe passage around them. The absolute devastation caused to shipping by these islands cannot be overstated. The whole group is a maritime graveyard, the final resting place for many fine ships which have been broken on these sinister rocks in centuries past.

The wholesale destruction of life and property was not helped by the islands being shown 10 miles further north of their true position on sea charts prior to 1750. How many vessels were lost because of the perpetuation of this historical inaccuracy is uncertain, but if this hazard did not drive many a good ship to its end, there were always the currents; vicious swirling currents which continuously sucked vessels on to the hidden granite teeth scattered throughout Scilly. With such a combination of factors against him, it was indeed a brave master who would take his vessel into the unlit waters around this area.

It was amongst these islands, too, that the Royal Navy experienced its worst peacetime disaster. In October 1707 a Naval Squadron was returning from the Mediterranean commanded by Admiral Sir Cloudesley Shovel. During the approach to Scilly an Atlantic gale of typical severity blew up, sweeping all but two of the frigates on to the ledges and reefs which litter the western seaboard of the islands. In a few short hours of 22nd October over 2,000 men perished amongst those dreadful rocks, while Sir Cloudesley Shovel was fortunate to be washed ashore in Porthellick Cove on St Mary's. He was found by the inhabitants who promptly slew him with a blow to the head in order to take possession of an emerald ring on the luckless Admiral's finger.

The dangers of these western reefs were well known. An early hydrographical book spoke of, "…many rocks terrible to behold…with the seas alternately flying over them in white sheets or fleeces of that element," while references to, "the dogs of Scilly" left visitors in no doubt of the terrible consequences which could result from navigational errors in this area. The setting up of a coal fired beacon on St Agnes in 1680, while of assistance to night travel amongst the islands, had little effect on the perils encountered while navigating through the rocks to the west of the main group. After the particularly distressing loss of the schooner *Douro* which struck on the island of Crebawethan on 28th January 1843 with the loss of all aboard, action was taken by Trinity House to survey the Western Rocks with the intention to subsequently erect a lighthouse. This time the light was to be in amongst the danger where it would be of the utmost benefit.

The site chosen could not have been more exposed or more dangerous. The lighthouse was to be erected on the outermost rock of the Scilly Islands, the western outpost of the British Isles, on a particularly evil rock which had sent many ships and their crews to a watery grave. It was called Bishop Rock, a name which has now been taken by one of the most famous and prestigious rock lights in the world. It is said that no other lighthouse exists on a more exposed site than this.

Left: An unusual artist's impression of what the Bishop Rock actually looks like under the water. It appears to be a pinnacle of rock that rises almost sheer from the floor of the English Channel. Inset is a view of the rock with its lighthouse.

Opposite: Bishop Rock lighthouse – a magnificent tower of granite marking the south western extremity of Britain (*Steven Winter*)

Right: At low water a surprising amount of Bishop Rock is visible. The red colour just above water level is not the colour of the rock, but a species of seaweed that thrives in the clear waters of Scilly.
(Richard Knights)

The exact derivation of this name is unclear. There exists a tale from the 17th century of the wreck of an entire fleet of merchant ships which struck this and neighbouring rocks, the only survivors having the names Bishop and Clerk, who were then honoured by their names being given to two of the rocks on which they so nearly perished. Alternatively, and more widespread, is the belief that the rock was so called because of its supposed resemblance to a bishop's mitre. Whatever its derivation, there is certainly nothing Christian about the way this pinnacle of granite claimed its victims. Over 150 ft long by 52 ft wide, it was completely covered by spring tides, becoming an invisible trap for a helpless vessel. This was no lengthy shoal, easily identified by foaming water breaking on its crest, but an isolated ridge of rock rising sheer from 120 feet and exposed to waves which had travelled across 3,600 miles of open Atlantic.

The force of water thrown upon this rock during storms is difficult to comprehend. How many people who are not intimately connected with the sea, for instance, can visualise an iron column weighing 3 tons being tossed 20 ft higher on to the rock, or an anvil of 1½ cwt being wrenched out of a hole 3½ ft deep, or the disappearance of an iron cradle suspended by a ½ in chain at the height of 115 ft above the level of the sea. All these incidents, together with countless 'minor' examples took place on the Bishop Rock. When the Atlantic rises in fury there is no force on earth with such awesome power, a power which was to show itself time and time again during the construction of the Bishop Rock lighthouses.

By 15th April 1847, all was ready for a start to be made on the Bishop. Working to the plans of James Walker, then engineer in chief to Trinity House, was the eminent Nicholas Douglass, assisted by his son James who was later to construct the present Eddystone tower. At this particular time it was not thought possible that the reef could provide sufficient area for the considerable base a masonry tower on this site would have required. Walker had therefore planned an open-piled structure with legs of metal, on top of which rested the accommodation and lantern. It was a design Walker would have seen at the Smalls where his stone tower replaced the wooden-stilted light in 1861. All the legs were of cast-iron with wrought-iron cores, the centre column being 3 ft 6 ins in diameter and hollow to provide access to the interior through a door 8 ft up this column. All legs were securely sunk and bolted into the rock and strengthened with cross members and ties between individual legs and the centre column. In four working seasons the tower was complete, except for the actual lantern apparatus which had been designed to have a range of 30 miles. This was due to be installed at the earliest opportunity in the following year which meant that during the winter of 1849 there was a lighthouse costing £12,000 with no light standing on the Bishop Rock.

It was far taller and slimmer than the structure on the Smalls, 120 ft between wind vane and rock, yet the principle it was designed to was exactly the same – the breakers, instead of meeting the solid resistance of masonry, would roll through the legs of the tower unhindered and therefore leave the light undamaged. This idea worked reasonably well on the Smalls, a site where the rock was always above the level of a normal sea, yet here on the Bishop it was a different matter entirely. On the night of 5th February 1850 the Atlantic rose in tumult, sending wave after massive wave crashing against the Bishop Rock. The legs of the tower snapped as if they were matchsticks, leaving only the fractured remains of the iron piles to mark the position of Walker's structure. He had sadly underestimated the exposure of the site and capabilities of an ocean in fury and had paid dearly for his mistake. A local newspaper, the *West Briton* reported on 15th February that:

The massive pillars and apparatus erected during the last three summers on the Bishop rock at a vast expense, were entirely washed down in the storm of Tuesday night last week...The pillars are broken off, some at the base, others at two, three, four and five feet from the foundation, evidently proving that the pillars were not sufficiently strong...

Below: A comparison of the three Bishop Rock lighthouses.

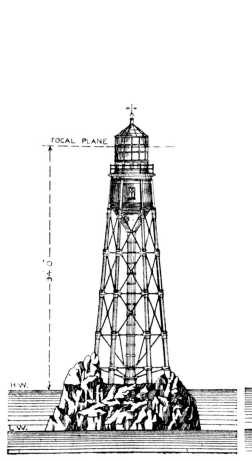

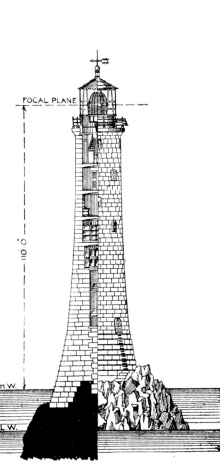

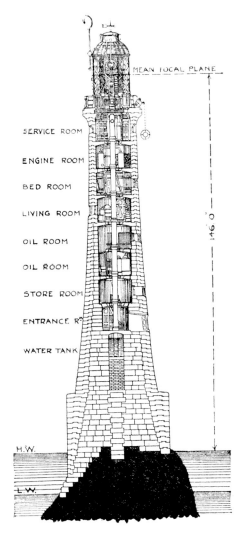

Bishop Rock 1
built 1847-50 by
N. & J. Douglass
(never operational)

Bishop Rock II
built 1852-58 by
J. & N Douglass

Bishop Rock III
encased Bishop Rock II by
W. T. Douglass 1883-87
(helideck added 1976, automated 1992)

In his account of The Bishop Rock Lighthouses William Tregarthen Douglass (the son of James), explained his views on the failure of the structure.

The disaster was doubtless due to the heavy seas sweeping over the rock at an elevation considerably higher than had been estimated, and thus striking with their full force the larger surface of the upper portion of the structure, intended for the residence of the lightkeepers, and for the storage of their provisions, fuel, etc.

These first attempts of designer and engineer must be regarded as nothing less than a spectacular failure and served as a salutary reminder to Walker and Douglass should they become complacent in the future. The one fortunate thing which can be said about this disaster is that the light was not actually functioning, and therefore manned. Had this been the case a repeat of the tragedy on Winstanley's Eddystone would have occurred.

Unperturbed by the failure of his first structure, Trinity House did not lose confidence in their engineer but instructed him to proceed with further designs for the same site. A year later Walker and Douglass were back in Scilly and hard at work on their second attempt, this time in stone. The base for the construction work was again on the island of Rosevear, 2 miles to the east of the Bishop, and in such a position that the site was clearly visible from the island. This was an invaluable asset, enabling the state of the surf around the Bishop to be estimated at a glance which made it possible to take advantage of every opportunity for landing.

It was not a particularly pleasant existence on Rosevear. As with all rock light construction the men were more often marooned at their base than working on the rock. When rough weather prevented landing it also prevented delivery of stores and materials to the base. Often the labourers were forced to resort to self sufficiency, living off what they could glean from the island and sea; fish, sea-birds' eggs, a few vegetables or even raw limpets. Although they were not confined to a floating barrack, life could still be particularly miserable on Rosevear.

Right: Nicholas Douglass' tower of 1849 designed by James Walker was the first Bishop Rock lighthouse. It had an open framework of legs, similar to Whiteside's structure on the Smalls. Before the lantern could be installed it was washed away by mountainous seas on 5th February 1850.

Nicholas Douglass was now a man of 52 years, whose agility must have been decreasing. It was obviously not practical for such a man to leap frequently from pitching boats on to wave-swept rocks and so the bulk of the construction work was now left to his son James while Nicholas acted as Superintendent. It was an arrangement which worked admirably well. James Douglass was a skilled and caring man who shared the same hardships and privations on Rosevear and the Bishop as his men, and in this way gained their respect and confidence.

The granite for the new light started arriving from quarries at Lamorna and Carnsew in Cornwall during 1852. They were dressed in Scilly on the adjacent St Mary's before travelling to Rosevear and the Bishop aboard the tender *Billow*, and later in a convoy of barges towed by a steam tug fittingly named *Bishop*.

The site for the base of the tower was anything but level, sloping sharply as it did to the north-west. This meant that the lowest courses of stones would not be complete until sufficient height had been gained for complete circular courses to be laid. The lowest stone, in fact, was planned for a position 1 ft below the lowest spring tides, for only in this way could the greatest possible diameter of the base permitted by the area of the rock surface be obtained. The first stone of the project was actually part of the fifth course and was laid on 14th July 1852.

Douglass and his men waited for what seemed an eternity to lay the lowest block of the 34 ft diameter base, for it was not until 30th July that it could be put to rest.

By the end of 1852, 44 stones – each dovetailed and keyed to one another – stood proud of the rock. Work progressed steadily, although much interrupted by inclement weather and gales. Even when work was possible the waves frequently swept the workings, covering the men at lower levels. A lookout was posted to give advance warnings of these combers, in which event the men clung desperately to iron stanchions, the rock, or each other as the huge weight of water deluged them. Occasionally, individuals would be swept into the foaming waters around the rock, but there were always plenty of volunteers to rescue their colleagues, often it was James Douglass who would fling himself into the icy waters after his men. The work was always fraught with such risks yet throughout the time it took to complete the lighthouse there was no loss of life or serious injury to anyone involved.

It took six gruelling years of toil and deprivation before the tower grew to its full height of 147 ft, 110 ft of which were above the high water mark. The first 45 ft were totally solid and the walls varied in thickness from 2 ft to 4 ft 9 ins. Its appearance was not dissimilar to Smeaton's Eddystone although its proportions were rather more grand. The lantern, 14 ft in diameter and 28 ft high, was installed and a fixed white light visible for 14 miles was first seen on the night of 1st September 1858, shining bright and clear across the open wastes of Atlantic for which many a mariner gave grateful thanks. The final account for lighthouse and buildings on Rosevear revealed that the price for lighting this evil spot was £36,559.

Much praise was lavished on the new structure. Prince Albert, addressing the Elder Brethren in 1858 considered the tower to be, "…a triumph of engineering skill and perseverance," while a visit to the site by the Royal Commissioners on 8th July 1859 left them in no doubt that this lighthouse was, "…magnificent, and perhaps the most exposed in the world."

However magnificent they found it, this second attempt at lighting the Bishop Rock, which at least produced a functioning beacon, was not to be entirely successful. For the reasons behind this we must look again at the quotation from the Royal Commissioners. The exposure of any particular lighthouse is not necessarily dependent upon how far away it is from the nearest landfall. What is a more realistic way of looking at this is to determine the length of unobstructed sea which wind and waves could pass over before they reach the lighthouse. This distance is usually referred to as the 'fetch'. It has already been mentioned that there are few, if any, lighthouse sites which have a fetch that can equal 3,600 miles of wild Atlantic that lie to the south-west of the Bishop Rock. After a journey of that distance, backed by a tearing gale, the pressure with which waves are driven against solid rock is staggering. Pressures in excess of 7,000 lb per sq ft have been recorded on the Bishop.

Although the first stone tower weighed 2,500 tons it was by no means insensitive to the shuddering blows of the mountainous Atlantic breakers which rolled up the sides of the masonry finally to spend themselves above the height of the lantern. During such gales the whole structure vibrated alarmingly, books were thrown from shelves, ⅝ in plate glass windows were smashed, allowing the sea into the rooms of the tower which quickly flooded, while the prisms of the dioptric apparatus fractured under the impact.

Less than two years after its inauguration, on 30th January 1860, one of the worst storms in the history of these islands assailed the tower. The devastation was disturbing; the metal ladder to the entrance door that was bolted to the exterior of the tower was torn from its foundations and flung into oblivion, the actual granite blocks a few feet above the high water mark were split by the concentrated force of water, and what was probably the most disturbing fact of all, the 5 cwt fog bell suspended beneath the lantern gallery was wrenched

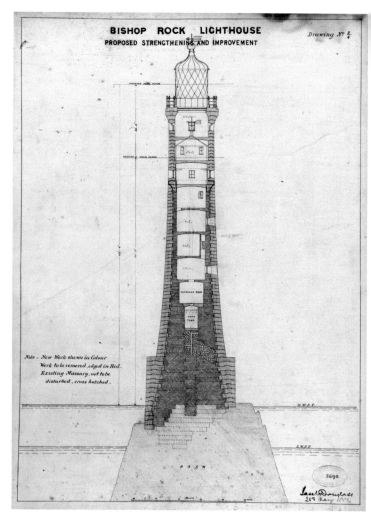

BISHOP ROCK LIGHTHOUSE.

Above: Two of the original Trinity House plans showing how the second Bishop Rock tower was to be strengthened and raised. The left hand plan is dated 29th November 1867 and the right hand plan is dated 26th May 1882. Both are signed by James Douglass. *(Trinity House)*

from its mountings and smashed into fragments on the rocks below. (One of these pieces was subsequently found in a cleft of the rock and is now preserved in Trinity House, London.) The waves rolled effortlessly up the sides of the masonry to spectacular heights, reflecting the light from the lantern back into the eyes of the terrified keepers. A flag of distress was flown from the gallery (the flag pole had met the same fate as the fog bell), but it was not until 5th February that a Trinity House inspector could land to view the damage which was later estimated at £1,000.

This was not an isolated incident. In an area where over 30 gales a year are not uncommon, a recurrence of such wild scenes was inevitable. On 20th April 1874 wind and waves combined in a fury that cracked lantern panes and shattered the lens in 30 places. Blow after blow rocked the tower, clawing out the cement between the blocks so that salt water found its way inside the building once again. All but two of the spare glass tubes of the Argand burners were thrown to the floor in the lantern room. The lantern could not function without these so the two intact survivors were carefully wrapped in cloth and preciously guarded, while all the time putty from the lantern glazing was being shaken loose and came cascading down. The keepers reported they could watch the waves passing the kitchen window, 70 ft up the tower, and feared for their lives with every sickening wave blow. When the seas eventually settled the effects could be examined. Sand from the ocean floor was found on the lantern gallery and once again cracks had appeared in many of the lower courses.

As a result of this, the tower was surveyed by a Trinity House engineer who recommended strengthening of the structure from top to bottom by bolting heavy iron ties to the internal surface of the walls and connecting them through the floors. This was duly completed but further storms in 1881 caused ½ cwt chunks of granite to flake from the exterior blocks, calling into question the success of the earlier measure.

In that same year Nicholas Douglass advised Trinity House that even more drastic measures were required. He proposed that the existing structure should be rebuilt by completely encasing it with an additional layer of masonry, dovetailed horizontally and vertically from the foundation courses to the service room floor, above which point the existing tower would be completely removed and four new rooms added. Also, around the base a cylindrical core of masonry was required (as found at the present Eddystone tower), which would act as a landing platform and help to reduce the force of water breaking against the base of the tower and so prevent waves climbing up it. At the same time, in an effort to reduce the number of wrecks in foggy conditions, Trinity House were keen to improve the illuminating equipment and fog signal. The result of these comprehensive modifications would hopefully put the tower, "…in the foremost rank of rock lighthouses."

Here indeed was an inspiring task for its executor, for not only was the old tower to be heightened and encased, all such work was to be done while a light was maintained throughout the period. Its completion set a new standard in civil engineering. The person to whom this mammoth task was entrusted was William Tregarthen Douglass, the son of James Douglass, who started his labours in 1883 when much of the equipment used in the construction of the new Eddystone tower was moved to a base on St Mary's. This included the 130-ton, twin-screwed steam tug *Hercules* which had proved invaluable at the Eddystone.

Particular attention was paid to the base of the casing as this had been shown to be the area of greatest stress and damage on the original tower. A cylindrical column of stone, 41 ft in diameter and 40 ft high, was raised around the taper of the older light, terminating in a landing platform or 'set-off'. It was necessary to extend the lowest courses for this even further beneath low water, and every block up to the twentieth course was not only dovetailed to the existing stones and its neighbours, but also bolted with 1½ in bolts into the block below, thus reducing the chance of displacement before the mortar had set hard. Furthermore, below the twenty-fifth course, both horizontal and vertical joints broke with those in the older tower. The first and lowest block was laid on 25th May 1883.

All the stones came from the Eddystone Granite Quarries at Delank, Cornwall. They were, following standard procedure, dressed and shaped in the workyards on St Mary's, before being loaded aboard the *Hercules* for the 7-mile journey to the Bishop. By the winter of 1883, 57 blocks had been set – nine complete courses.

This was not a spectacular rate of progress for many reasons. Firstly, the blocks not only had to fit one another, they also had to fit the dovetails being hewn out of the existing tower. This task of preparing the outside faces of the old blocks was proving a tedious business as great accuracy was required. Also, at these lower levels, where a large proportion of new blocks were to be laid, the men were in constant danger of being washed off the rock. To reduce such accidents Douglass fixed a stout chain around the circumference of the tower to which life lines were attached. When a wave broke over the site the men clung to these lines while the waves washed over them. The higher the blocks rose, the less dangerous this task became.

During 1884 the weather was kinder to Douglass enabling more frequent landings to be attempted. When work stopped for the winter on 19th November, 457 stones had been laid constituting 22 complete courses. Once the work had risen above the mid-tide level, about 30 men were lodged in the actual tower itself. Their job was to continue the external preparation of the old tower even though it was not always possible for men to land and set new blocks because of the state of the sea. The masons lodged on the rock carried out their work from wooden foot boards resting on brackets lodged in between the blocks. Beneath them was a manilla rope net in case of accident. This system gave the men complete faith in their safety as they persevered with their exacting task.

By 9th December 1885, 54 courses were complete, and by 11th May 1886 the dovetailed casing was level with the lantern of the old lighthouse. Up to this point all the new blocks had been set in position using a crane on the lantern gallery that was capable of traversing around it and therefore covering all sectors of the casing. Once the lantern was reached it was immediately dismantled, together with the walls of the old service room, and a wrought-iron mast 40 ft high placed in the middle of the service room floor. This was to be the basis of a crane for all further work and on top of it was affixed a Trinity House floating-light lantern with a powerful double flashing light of 16,000 candle-power intensity produced by six burners. The crane, lantern and illuminating apparatus were all installed without any break in the nightly warning the lighthouse gave, and were raised by hydraulic jacks to successive positions as the work progressed.

Work resumed on 26th June 1886 with much speed. By 30th August of the same year a further 28 courses, numbers 63 to 91, were set at the rate of one or two courses a day. This was a far cry from the slow laborious work involved in the lowest courses. The sea, however, never let them forget what a tenuous existence they led by clinging to a growing stone column. One north-westerly gale was powerful enough to rip the heavy steam winch – which was protected by masonry and bolted to the 'set-off' 22ft above high water – completely off the lighthouse and toss it into the wild sea, leaving only its mounting bolts and part of its bed plate behind.

Three and a half years had elapsed since the first stone of the casing was laid. Now, 3,220 tons were in position, giving a stronger second skin to the earlier Douglass tower. A new, improved lantern was fitted on top of the taller masonry, consisting of 20 burners arranged in two tiers of ten, and capable of generating 230,000 candle-power. This lantern, now revolving around an electric light bulb, was 14 ft in diameter and 15 ft high. As the height of this lantern had been increased to 144 ft above the sea, a corresponding increase in the visibility of its double white flash every 15 seconds was also noted, which could now be picked up from a distance of 29 miles. The new machinery was of such a weight it was found impossible for it to be rotated by clockwork, as had previously been the case, so a 1½ hp motor was installed to meet this deficiency. The old fog bell was replaced by a gun cotton explosive signal which proved more efficient.

The new light was brought into service on 25th October 1887 and cost an estimated £64,889. Following in the footsteps of his father and grandfather before him, William Tregarthen Douglass had maintained the standards of skill, courage and craftsmanship that had become associated with the name Douglass. His work on the Bishop Rock, with its intricacies of dovetailing, heightening and strengthening, while all the time maintaining a light, was a precedent in the field of lighthouse engineering. No such work had been attempted before and the fact that it remains today, exactly as Douglass had finished it, is testament to the precision and skill of its builder. In appearance it bears all the hallmarks of a Douglass tower, being almost the twin of the newest Eddystone. Only the slightly different shaped lantern roof and taller cylindrical base distinguish them.

The effects of the two Bishop Rock lighthouses were dramatic. The menacing reefs, once plunged into inky blackness every sunset, were now pinpointed by a brilliant spoke of white light. Yet even with such safeguards there was still one evil the effects of which no lighthouse can ever totally negate – fog. Here it is perhaps worth recounting the tale of the final hours for a vessel that was lost not ¾ mile from the first lighthouse in one of the thick, swirling fogs that descend upon the sea in this area with such alarming rapidity.

The year is 1875. Douglass' first lighthouse had stood for all but 17 years, and although reeling from the effects of heavy seas, was still serviceable and continued to light the night sky around Scilly. Early in May, the *Schiller*, a 3,421-ton mail steamer belonging to the Eagle Lines, was approaching the island en-route for Hamburg after a trouble-free crossing from New York.

On board were a crew of 118 and 254 passengers. On 6th May, with the sun setting astern of the vessel, most of the passengers were settling down to dinner. It had been an uneventful voyage with calm seas and clear skies, and for a handful of passengers due to disembark at Plymouth it was their last night on board. The celebrations had just commenced when, unbeknown to the revellers below, the weather took a turn for the worse. Thick, swirling fog banks closed around the speeding vessel as she approached Scilly. The captain was preparing to be guest of honour at a dinner given by some of the passengers when he was told of the change in weather. At this time prestige on the thriving trans-Atlantic run was judged, not on the standard of service or luxury of the fittings on board, but on the time taken to complete the crossing, so Captain Thomas was unwilling to reduce speed until the deteriorating conditions and thickening mists made it folly not to do so.

Also, unknown to anyone on the bridge, the unusual currents around Scilly had put the *Schiller* off her true course, an error which could not be rectified because of the poor visibility. This phenomenon, together with the ludicrous speed with which the *Schiller* was driven through the fog, combined to produce a fatal catastrophe. At two minutes to ten the *Schiller* struck the Retarrier Ledge and gashed herself open along her length. Towering waves continued the destruction of this once-proud vessel and by the morning the terrible fate of the *Schiller* was all too apparent. Barely recognisable minus funnels, bridge and masts, the hulk of the vessel lay broken where it struck. Out of a total complement of 372 only 37 escaped with their lives to recount the full horror of the previous night. Tales were told of how periodically the mists would part to give a clear view of the Bishop Rock lighthouse, fog bell tolling, brilliantly illuminating the scene of a crippled ship being pounded mercilessly by heaving seas. The obsession of a captain for a fast crossing at the expense of all else had contributed to one of the most tragic maritime disasters of the century. The keepers of the lighthouse, although well aware of the terrible events taking place only a few hundred yards from where they stood, were powerless to help but could only watch the morbid scenes unfolding before them. One of the keepers that night was moved to record the full horror in a letter to his wife, part of which reads as follows:

At 11-35 pm, William Mortimer came running down and called to me, and said he could see a vessel on the rocks. I jumped up and went out on the parapet without stopping to dress, and saw the masthead and starboard light of a large steamer. She was burning blue lights and firing off guns and rockets. She seemed to be sinking. The last gun was fired at 1-30 on the 8th inst. I relieved G. Gould at four am.

Fog again raised at six am, and I could then just see the topmast of the vessel out of the water. We could count about 26 people in the rigging. I could see one lady in the lee side of the rigging with two males by her. She was in a sitting posture, I should think lashed. It was a dreadful sight. At about seven am the mast fell, and I supposed every one perished, but I still hope a few or some might have been saved.

On Sunday three bodies floated past us, and this afternoon more have passed close to us. No one knows what was felt in this house by all hands to see so many of our dear fellow-creatures suffering and dying so near to us. Their sufferings must have been severe, for it was a cold drizzling rain all night, wind W.S.W. I think

Above: Because of the violence of winter storms and the damage they caused to the second Bishop Rock lighthouse, it was decided to encase the tower in a layer of masonry and increase its height. It took William Tregarthen Douglass, the engineer in charge, from 1883–87. This photograph from 1885 shows the old lighthouse disappearing inside its new encasement. *(F. E. Gibson)*

you had better take this letter, together with my compliments, to Mr John Banfield, Lloyd's agent.

I remain your affectionate husband,

JAMES DANIEL.

The tale of the *Schiller* is still retold today as a warning that even the warning devices of a lighthouse can be rendered useless during fog.

Tales of disappearing lighthouse keepers are notable for their scarcity, so that when one does occur it tends to make the headlines of the day. Readers will recall the events at the Smalls lighthouse earlier in the book; not exactly a disappearing lighthouse keeper although his death did have far-reaching effects for the whole of the British lighthouse system. There will be another incident of disappearing keepers in a later chapter, but first I must report the strange incident that happened at Bishop Rock in 1898 – one that has never been satisfactorily explained. *The Cornishman* newspaper of 29th December 1898 carried the following report;

THE SAD EVENT AT THE BISHOP ROCK LIGHTHOUSE

General excitement prevailed at St Mary's Island of Scilly, on Monday evening at 8 o'clock, when it became known that distress signals had been made from the Bishop lighthouse to St. Agnes. That excitement was intensified when it was ascertained that no relief was required till the morning. Surmises were rife as to the nature of the distress. It was evident that something had happened and, as help was not immediately required, it was feared that death had made an inroad among the little staff of lightkeepers.

Early the next morning a relief was made, when it was ascertained that Mr Ball, the principal, was missing – how or by what means his comrades knew not.

It appears that, at 3.30 on Monday afternoon, he left the top of the tower and descended to the base for the purpose, it is thought, of taking an airing on the plat – a large piece of flat rock just at the base of the tower. He had been in the habit of doing so previously. No notice was taken of his absence until 4.30, the time for lighting up. It was Mr Ball's watch.

As he did not turn up at the appointed time, the other men became very anxious and instituted a search: but all to no effect. There was nothing left to show by what means he had disappeared. It was supposed that he either slipped off the rock or was seized with giddiness and fell into the water, as at the time there was no sea on to wash him off the plat.

Great sympathy is felt for Mrs Ball and family. Mrs Ball was at the Lizard at the time of the accident.

This is the fourth principal that has died within the last six years; the other three – Messrs Ranbridge, Nicholls and Brown died on shore.

A further notable incident, this time during the history of the strengthened tower, occurred on 22nd June 1901. A strong wind from the south drove the four-

Left: The BBC reporter Edward Ward was marooned in the lighthouse for four weeks longer than intended by heavy seas after his Christmas broadcast from the lighthouse in 1946. He's just visible leaving on a rope stretched between the tower and a waiting boat.
(F. E. Gibson)

masted *Falkland*, journeying between San Francisco and Falmouth with wheat, actually on to the Bishop Rock, causing her yard arms to scrape against the exterior of the lighthouse while the reef tore deep into her hull. The crippled vessel was washed clear to the north of the tower but was filling rapidly as the crew desperately tried to launch boats. Several lives, including that of the captain, were lost.

To detail all incidents, both minor and not so minor, involving the Bishop Rock lighthouses would take considerably more space than is available her. Suffice to say that lives too numerous to count have been spared by the construction of these structures and the warning they give. There is many a grateful mariner who owes his life to this particular lighthouse.

Older readers may recall the drama which centred on this lighthouse during the winter of 1946. The BBC had planned a Christmas broadcast from the station to illustrate the often lonely and unsocial existence of a lightkeeper, particularly at this time of year when families tend to unite as one. The broadcaster, Edward Ward, accompanied by a sound engineer, were duly landed and completed the scheduled programme. They had hoped to return within three days but the heaving seas around the Bishop kept them imprisoned for four miserable weeks. It was his unscheduled broadcasts that conveyed the true isolation of a lightkeeper's life far more vividly than a prepared script could ever do.

Over the years the Bishop Rock lighthouse, apart form its primary function of marking the reef on which it stands, has become an unofficial marker for the eastern end of the Blue Riband race – the 2,949 miles between the Ambrose light vessel outside New York harbour to the Bishop. The famous passenger liners of the recent past, household names such as *Queen Mary*, *Queen Elizabeth*, *United States* and *France* have all competed in this marathon run, the record still standing at 3 days 12 hours 12 minutes, set by the *United States* on 15th July 1952 – the last of the great ocean liners to do so. In 1998 the ferry *Catalonia* held the Blue Riband for a matter of weeks before the *Cat-Link V*, a 91-metre car and passenger catamaran made the crossing in 2 days, 20 hours and 9 minutes at a speed of 41.3 knots.

Finally, before this chapter closes, mention must be made of the recent technology to be found crowning the lighthouse. Trinity House had been well aware of the difficulty and uncertainty in effecting reliefs of the keepers on schedule, reliefs which involved dangling men and supplies on the end of a rope stretched between the lighthouse and attendant relief boat. A miscalculation by the skilled boat crew could result in the vessel gashing itself on the jagged rocks from which the tower rises, or a sudden wave slackening the rope between tower and boat could land the cargo, be it human or supplies, in the foaming waters around the granite teeth A near perfect day was required to effect a relief, always spectacular events and a popular tourist attraction during the summer months. Delay meant frustration for both those on the light and the families on shore.

To this end, during 1976 a helideck was erected above the lantern to provide a reliable and efficient method for the relief of keepers and delivery of stores that was not dependent upon the state of the sea. It was undoubtedly a technological advance of the highest order and won for its designer the Structural Steel Design Award, yet at the same time one wonders if the erection of this ironmongery has not spoilt the graceful lines of Douglass' tower? It has since become a standard feature at all of Trinity House's remote stations where landings are uncertain.

It was certainly vital to the successful automation of the station in the early 1990s. Now the only access to the station is through the helideck! Automation meant modification of one of the largest optics in the Trinity House service – a huge Hyper-Radial 1st Order Bi-form Asymmetrical Dioptric unit, a description almost a big as the optic itself. The top tier of this

Above: The massive base of Bishop Rock lighthouse from an approaching boat. When conditions were calm enough for a boat relief to get as close as this, there was still a exhausting climb into the tower. Various photographs show that there are about 52 near-vertical steps before the entrance door is reached.
(Joy Adcock)

giant was removed in order to fit an emergency light. Instead of the original 1500W filament lamps two 400W metal halide lamps mounted below this in an automatic lamp changer now form the main light with a range of 24 nautical miles. Its character remains unchanged, although the fog signal and fog detector, in common with normal automation practice have become electric instead of compressed air.

An interesting development of the automation process at Bishop Rock is the introduction for the first time at a Trinity House station of what is known as the 'cycle charge system'. The high power requirements at this station ruled out the use of solar panels for charging the batteries (a practice becoming increasingly common at automatic lighthouses) and this would normally require the newly installed diesel generators to run continuously, with 6 months between maintenance visits. The cycle charge system involves the diesel generators only running when the state of the battery charge drops to a certain level – the batteries providing all the power requirements of the station. It's a system that reduces fuel costs, the frequency of fuel replenishment flights and the interval between maintenance visits. The fuel for Bishop Rock now needs replenishing only every 18 months. Full automatic operation was declared on 15th December 1992 when the last keeper was withdrawn, thus ending 134 years of manned operation. The lighthouse is now monitored by telemetry from the Trinity House Operations Control Centre at Harwich.

Figures for Bishop Rock lighthouse released by Trinity House clearly indicate the financial benefits of automation. Before automation it cost £191,000 per annum to run the station. The automation process itself cost £450,000 but the annual running costs are now estimated at just £96,000 – a saving of £95,000 per annum. It is estimated that for the 15 year life of the new equipment a saving of £975,000 will be made. The lifeless Bishop Rock lighthouse still remains the first sight of Britain for many trans-Atlantic voyagers. Its double white flash every 15 seconds welcomes and warns vessels approaching our shore. Its familiar profile, breaking the horizon, is a monumental example of civil engineering on one of the most inhospitable sites which could be imagined. It once again brought prestige and glory to the Douglass family, although the achievements of these remarkable men hardly needed more acclaim. It stands sentinel over the maritime graveyard amongst the islands of Scilly where it has done so much to reduce the effects of the once-feared rocks and ledges.

Opposite: A stunning shot of the outside of the lighthouse after automation. The windows of the Store Room door have had heavy metal plates bolted over them and the entrance doors below will probably never open again. The iron bolts that hold the copper lightning conductors in their channel are rusting, and most interesting of all is that the outer surface of the granite blocks appears to be flaking off!
(Joy Adcock)

Left: The roof of the original lantern has been removed and scaffolding erected around it prior to the erection of a helideck in 1976.
(MOD Hydrographic Department)

THE WIDESPREAD nature of the British Isles has always caused map makers to scratch their heads when trying to put all of the British 'isles' on the same map. Frequently they gave up – trying to include all the far-flung island 'groups' on the same map as the 'mainland' meant there was far too much blue ocean surrounding the land where the maximum detail was required. Groups such as the Channel Islands, Scilly Isles, or Shetland were very often missed off early maps altogether, such was their distance from the English or Scottish coastline. The Inner and Outer Hebrides and Orkney fared rather better – their closeness to the 'mainland' ensured their inclusion on even the earliest maps.

Their answer to this tricky problem – how to include all parts of an oddly shaped United Kingdom on the same map – was to put the outlying parts that didn't fit neatly into a rectangular 'map' shape into boxes in the corners of the main map. In this way, they could justifiably claim to be including all the constituent parts of the British Isles on one map to show the greatest detail. It's a system American map-makers have used to get over a similar problem they have with Alaska and Hawaii.

Although this 'inset' method is a perfectly valid technique for map-makers, it does have one big disadvantage. It doesn't give a true picture as to the exact position of the area included within the box, even though lines of latitude and longitude may be included within the 'box'. It's difficult to try and imagine the position of an inset area in relation to the rest of the map when you don't see the area as a whole. Nowhere is this better illustrated than in the case of the Shetland, Britain's northernmost island group.

The inhabitants of Shetland are now well used to seeing their islands appearing in a box in a corner of a British Isles map, such is their distance from the nearest 'mainland'. It's a problem perpetuated by television weather forecasts, school atlases, road maps and countless other examples. It's unfortunate because the result is that British citizens rarely get a true picture as to the exact location of these historic isles. Examination of a map of northern Europe gives a clearer picture.

The Shetland group is an archipelago of islands stretching for over 70 miles in the North Atlantic, about 125 miles north-east from John O'Groats. The capital of the group, Lerwick, is located at 60°10′N. This places it further north than Oslo, Stockholm and Moscow and at about the same latitude as Helsinki. It is also nearer to the capitals of Norway and Denmark than to London, and a sea voyage to Bergen in Norway would take less time than a voyage to Edinburgh, the capital of Scotland. Only by examining a map of northern Europe in this way can you get an impression of exactly how far north these lonely isles are in relation to the rest of the United Kingdom. It's no surprise therefore that the Shetland group are often referred to as the 'Northern Isles'.

The subject of this chapter is found on one of the northern isles, almost the most northerly of the Northern Isles and within half a mile of being the most northerly point of the United Kingdom. It's a lonely, wave-lashed, wind-swept lump of rock that holds a unique place in the British lighthouse service. Muckle Flugga is Britain's most northerly lighthouse. Its construction in such northerly waters tested the skills of David and Thomas Stevenson to the limit. It is also unique in lighthouse history because its construction was the direct result of a war.

The Crimean War of 1853–56 was one of the long series of wars between Russia and Turkey, but it differed from the others as a result of British and French involvement. Tensions increased during 1853 and long negotiations, in which the British backed the Turks, failed to resolve the problems. Public opinion in Britain, France, Turkey, and Russia contributed to the rigidity of the respective governments. The Russians then destroyed the Turkish fleet at Sinope

Opposite: A dramatic view of Muckle Flugga and its lighthouse showing the almost sheer sides of the rock. Everything needed for the construction of the lighthouse had to be carried to the summit by the men building it. The original steps hacked into the eastern face have been replaced by the modern metal steps seen in this view.

(Northern Lighthouse Board)

Above: Muckle Flugga
is the northernmost
lighthouse in the
United Kingdom. In the
foreground is Out Stack
– the last piece of the UK
before the Arctic Circle!
(Author)

on 3rd November 1853. This so-called Sinope massacre raised war fever in Britain and France, and these nations declared war on Russia on 28th March 1854. Part of the British government's strategy in this war was to blockade the White Sea on Russia's northern coastline to prevent the Russian fleet sailing from Murmansk and Arkangel for the Mediterranean. Unfortunately, navigation aids along the route to and from the White Sea were practically non-existent, particularly around the treacherous waters of Orkney and Shetland. In 1854 there were only three lighthouses amongst the entire two island groups.

The Admiralty in London decided that further navigation aids were required to ensure safe passage for the British fleet to and from Arctic waters along the north and east coasts of Shetland. They were perhaps mindful of the events of a night in 1811 when three warships, including the 98-gun *St George* and the 74-gun *Defence* were lost in a North Sea 'hurricane' that engulfed the flotilla returning from the Baltic. In this one single night over 2,000 seamen were drowned – over twice the number lost in the entire Battle of Trafalgar. Accordingly, they advised the Board of Trade, who since The Merchant Shipping Act of 1853 were the official government 'department' that controlled the workings of the three British Lighthouse authorities, to have two further beacons operational "if practicable" by October 1854. They in turn informed Trinity House who at the time had the power to direct the Northern Lighthouse Board to erect, or even dismantle, navigational aids in Scottish waters.

This was an incredible task. To survey and build two new lighthouses around the wild Shetland coastline would be a daunting task in itself, but to accomplish it within a matter of months seemed an impossible request. However, it's a sad illustration of the fact that in times of conflict Man can channel all his energies into accomplishing tasks normally considered impossible in such a short time scale. David Stevenson, the NLB's Chief Engineer, in whose lap had landed the task of constructing these lights later remarked, "It was no joke to have the responsibility of having lights ready for our Fleet after the Government had given their opinion that these lights were necessary and should be exhibited before winter set in."

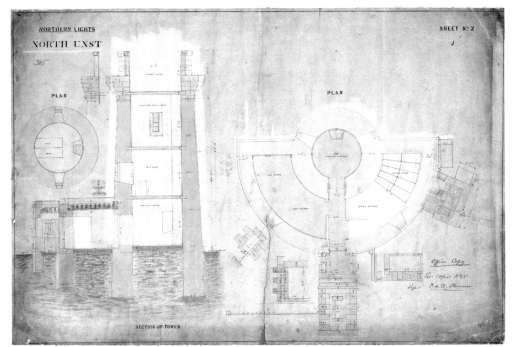

Left: The original hand-painted plans for Muckle Flugga lighthouse dated 5th April 1855 and signed by 'D & T Stevenson'. They show that the foundations for the tower go down 10 feet into the solid rock and are 4ft 6ins thick at the base. (*Northern Lighthouse Board*)

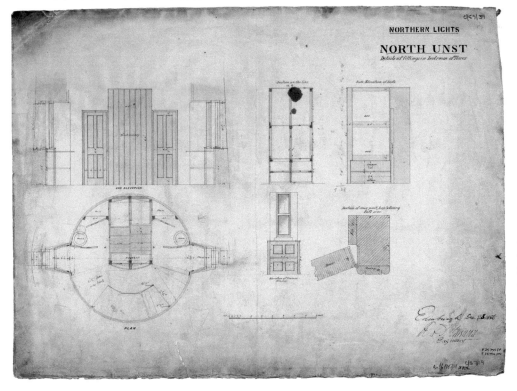

Left: The incredible detail of the Muckle Flugga plans is shown in this example for the bedroom fittings. The carpenters building the lighthouse would use these plans to show them how to fit a hinge to a door frame! (*Northern Lighthouse Board*)

Two sites were identified; the island of Whalsay, a couple of miles east of the Shetland Mainland, and somewhere on or near the island of Unst in the far north. The Commissioners had already started work on a light on Out Skerries, an outlying group further east of Whalsay, but the site of the new northern light was causing some problems. Two sites at the north-eastern and north-western extremities of Unst were being considered, along with one on a group of half a dozen uninhabited satellites called the Burrafirth Holms, half a mile off the very northern tip of Unst. One of the Burrafirth Holms had already attracted the attention of Alan Stevenson in 1850 who, when on a tour of possible lighthouse sites, had managed to

Right: A contemporary engraving of Muckle Flugga lighthouse (or North Unst as it was known then) – probably done by someone who had never seen it! Although the tower is too tall and the rock itself a little exaggerated, it does convey something of the lighthouse's dramatic isolation.

"spring ashore" on Muckle Flugga, the highest of the Holms, smash off a lump of its rock, before retreating to his boat again.

By February 1854 David Stevenson was on his way to Shetland to survey a site for the new lighthouse on Unst. The inhabitants were at first surprised to see him, being unaware of the declaration of war, and of the Commissioners intention to build a new lighthouse. Bad weather had prevented any communication with the rest of the country for many months. He set about surveying various possible locations for the proposed lighthouse, and during this time remarked that;

When surveying the several sites on which we landed I had a very favourable opportunity of witnessing the effects of the storms during the recent severe winter which had left their traces in unmistakable distinctness on every headland. These were particularly noticeable at North Unst where the deep water comes close to the rocks... and I came to the conclusion that what I had formerly considered as abnormal seas occurring at certain exposed places such as Whalsay were in point of fact normal on the whole of the north and east coast of Shetland.

The weather was sufficiently bad that he was unable to land on Muckle Flugga. He did, however, get close enough to form the opinion that it would be impossible to build a light there and proposed Lamba Ness in the north-east corner of Unst as the favoured site.

The Commissioners informed the Board of Trade of their engineer's decision, adding that they would regard it as, "culpable recklessness as regards the lives of the lightkeepers' to erect even a temporary lighthouse on the Flugga." This obviously didn't please the Board who promptly despatched a committee of Trinity House Elder Brethren to Unst to see the site for themselves. They landed on Muckle Flugga without difficulty on a quiet June day, decided that the building of a lighthouse on its crest would be "quite practicable" and made flattering remarks about how the Commissioner's "eminent engineer" would overcome the obvious difficulties. Both David Stevenson and the Commissioners pointed out to the Elder Brethren that Muckle Flugga on a flat calm day in midsummer was a very different prospect to when winter gales were being driven over its summit. Even so, they eventually capitulated and agreed to build a temporary lighthouse on Muckle Flugga, but would accept no responsibility for its successful completion or regular exhibition during the winter.

On a later visit at the end of July, Thomas Stevenson, David's brother, together with Alan Brebner, the son of a mason on the Bell Rock lighthouse, managed to pull into a sheltered gully below its eastern face and leap onto Muckle Flugga. They scrambled to its summit and ascertained that there was indeed just enough room to built a temporary beacon. The ascent of the treacherous face was difficult but not impossible, and was often compared to climbing an Alpine peak. What was going to be so unusual about this site was the fact that Muckle Flugga isn't a low lying half-submerged reef of the kind the Stevenson family had become accustomed to building on, but a huge razor backed lump of rock that rose sheer on both sides out of the north Atlantic to a height of 200 feet. The light was to be built on the summit of the rock and every single item required for its construction had to be manhandled to the summit on the backs of the men who were going to build it.

Left: An unusual view of Muckle Flugga taken from the NLB helicopter hovering below its western face. *(Author)*

Despite protests from the rock's owners, Lord Zetland and Edmonston of Buness, the Commissioners purchased Muckle Flugga and its northerly outlier, Out Stack, for £100, together with the right of communication with the rock from Herma Ness. Upon his return to Edinburgh, David Stevenson had to prepare his plans and put into operation the ordering of construction materials, lantern, optics, the sections of a temporary iron barrack, together with the hiring of boats and training of keepers. He had precious little time if the light was going to be operational within months. On 31st July 1854 a start was made when a temporary lighthouse and iron dwellings left Glasgow on board the lighthouse tender *Pharos*.

A few days later, under the supervision of Alan Brebner, a boatload of workmen edged their open boat into a narrow creek under the eastern face, leapt onto the slippery rocks and started to scale the almost sheer sides of this towering monster. Once its summit was gained,

Overleaf: A typical day at Muckle Flugga – grey with a strong northerly wind stirring up the sea below the lighthouse. In the distance shafts of sunlight break up the squally showers. *(Jean Guichard)*

Left: By taking down a radio mast there was just room to squeeze a helipad onto Muckle Flugga. The keepers in their orange survival suits watch the approach of the NLB helicopter. (*Author*)

life-lines were attached and a series of rough hewn steps were hacked into the face. Incredibly, a flat area for the light was cleared on the summit and was ready for building by the middle of September.

Here it should be remembered that everything, literally everything, required to construct the lighthouse had to be carried up the face of Muckle Flugga on the backs of the men building it. Over 120 tons of tools, provisions, blocks and tackles, sheerlegs, the sections of the temporary iron barrack, cement, and blocks of stone were all manhandled or hauled up the ferocious eastern face. Twenty masons and assorted labourers lived in the barrack on the rock, while sixteen other labourers made a daily journey from the shore base at Burrafirth. The tower was 21 feet high and topped with a cast iron lantern. Incredibly, it took just twenty-six days to construct this temporary light on the summit of Muckle Flugga which was lit for the first time on 11th October 1854. The iron barrack that had housed the labourers now served as accommodation for the keepers, its outside walls encased within a wall of protective rubble quarried on the rock and set with cement.

During that first winter Charles Barclay, one of Stevenson's foremen, was left on the rock with the keepers and a gang of men to complete the stairways being cut into the eastern face of the rock. Barclay was a seasoned lighthouse veteran, having worked at Skerryvore (where he lost a hand), Barra Head and other exposed lighthouse sites around Scotland. He was well used to seeing the effects of a sea in fury and not prone to exaggeration or distortion, so the Commissioners were rightly concerned by his reports of events at Muckle Flugga during storms in December and January. The lighthouse vessel *Pharos* was despatched to Muckle Flugga with spare lantern panes, lamp glasses and cement, and would return with Barclay. He related stories of how "sheets of white water" were driven over the rock during winter storms with spray being carried forty to fifty feet above the lantern. Waves, he reported, struck the lighthouse buildings, "with a dash" not to mention the earth, stones and grass that were found littered around the tower after a storm.

One December morning in particular the sea struck Muckle Flugga with such force that it climbed to the summit of the rock, smashed open the iron door to the dwelling house which weighed nearly a ton, and admitted a three-feet high wall of water which swirled around inside the house before retreating taking everything that wasn't firmly secured with it. Before the keepers had time to close the door, another huge wave surged in, followed moments later by a third which struck the roof with a deafening roar. The principal keeper, a man called Marchbanks, reported that such demonstrations of power by the sea were by no means uncommon. Waves frequently broke against the tower and dwellings, carrying earth and debris up to lantern height, or flooding the courtyard surrounding the light. Forty feet of stone 'dyke' had been carried away by the sea, along with six water casks. "We had not a dry part to sit down in, nor even a dry bed to rest upon at night," Marchbanks said.

Here it should be remembered that all these spectacular examples of a sea in fury were taking place at 200 feet above sea level – that's over 30 feet higher than the helidecks on top of Bishop Rock and Eddystone lighthouses, Trinity House's two tallest towers, and 50 feet higher than the top of Nelson's Column in Trafalgar Square! The seas that smash against the solid face of Muckle Flugga do so at the end of a journey of three thousand miles of open Atlantic, and when driven by hurricane force wind can climb to unimaginable heights. That first winter a daily watch was kept on the light from Herma Ness on Unst, but it was frequently obscured by curtains of rising spray and mountainous seas.

During the winter David Stevenson was called to London to discuss the situation at Muckle Flugga with the Board of Trade and Trinity House who were represented by James Walker. They all agreed that the buildings at the site were not free of risk, but that the risk to shipping would be greater if the light were not there. "While the buildings at Muckle Flugga could not be pronounced free from risk, that risk would not be greater or probably so great as would be incurred by shipping if the lighthouse were removed" was the official statement. The need for a permanent structure somewhere in the region of North Unst was agreed, although there was still disagreement as to its permanent location. The Commissioners still favoured Lamba Ness, but the Board of Trade and Trinity House favoured making the temporary light on Muckle Flugga permanent. There was considerable acrimony over this apparent stalemate which resulted at one point in March 1855 in the Commissioners asking the Board of Trade to relieve them of their duties in managing the Northern Lighthouses. They were reminded that only Parliament could do that, but when certain assurances and soothing words were given in response by the Board of Trade, the situation was calmed and nothing more became of the Commissioners request. Permission to build a permanent light on Muckle Flugga was granted in June 1855.

Left: The NLB helicopter changing keepers at Muckle Flugga *(Author)*

The permanent tower that Stevenson planned would a radical departure from others in such exposed positions in that it would be built of brick rather than stone blocks – "an untried experiment in marine engineering." Work started in April 1856 by sinking foundations ten feet into the summit of the rock. Over 100 men were employed in this season and their endeavours were helped somewhat by the construction of an inclined plane worked by a 10hp steam engine to haul materials up the face of Muckle Flugga. This device alone was calculated to save the work of at least sixteen men. In just forty working days all the component parts of the incline engine itself, its boiler and workmen's barracks had been landed, hauled to the summit and bolted to the rock. On North Unst, workshops and houses were being built.

Choosing bricks for the tower meant it was possible to land them from smaller open boats – sometimes by 'throwing' them from the open boat to the workmen standing on the landing – that could approach the rock in weather that would have made it impossible to land

heavy stone blocks or iron sections. In this way progress was rapid. The walls of the tower were 3½ feet thick and rose to give the tower a final height of 64 feet. At regular intervals during its construction waves climbed the rock and swirled around the men's feet, although to this day there have never been reports of waves penetrating the walls of the completed tower. David Stevenson decided to reduce the effect of such waves by building a circular protective wall around the lighthouse. While this offered some degree of protection it was still reported

Right: An unusual view of Muckle Flugga taken from almost directly above it in 1972. At this time one of the two radio masts is still visible to the north of the tower. At that time it was the tallest structure on the rock as it was considerably taller than the tower. It was subsequently dismantled and replaced by a helipad. (*Northern Lighthouse Board*)

that the keepers had to move around outside the tower on their hands and knees during high winds. The permanent structure on Muckle Flugga was lit on 1st January 1858 and was estimated to have cost £32,000.

In succeeding years Muckle Flugga has witnessed many storms of seismic fury, yet the light has continued in unbroken service. Every so often, the keepers report instances of spectacular wave damage such as smashed lantern panes, pathways being swept away, or the materials stored in the courtyard being carried off by waves. In one winter storm, seas were reported to have been breaking over the rock for 21 hours continuously, during which time they swept away one gateway pillar and loosened the other. Huge blocks of stone, two feet square were described as being, "rushed over the court[yard] as if they had been wood." There is even the story of a live fish being found in one of the summit pools by a lightkeeper.

On 20th March 1995, Muckle Flugga lost its keepers, as automation finally reached this northern outpost of Her Majesty's territories. Incredibly, and prior to automation, by dismantling a huge radio mast, there was room to squeeze a helideck onto the summit of the rock, just to the north of the lighthouse buildings and the surrounding wall. Before this, the keepers had to rely on exactly the same method of relief as the workmen who built the light: edging into the same narrow gully below the eastern face and leaping onto the slippery landing stage and steps before starting on the leg-aching climb to the summit.

It's only when you set foot on Muckle Flugga that you get a true feeling of its exposure and isolation. My first visit was on an April day when a boat relief would have been impossible, but a bumpy helicopter ride from Stromness in Orkney with the relief keeper in the NLB helicopter was possible. Standing on the helipad, white water was crashing against the rock below me, a biting wind was gusting from the west. I could feel droplets of spray on my face that had been driven up the west face of the rock. A brief detour in the helicopter to Out Stack, half a mile further north to photograph the very last piece of United Kingdom before the Arctic Circle, gives an even more vivid view of this wild rock. Sitting on the end of a ridge of saw-toothed rocks, the tiny white tower is in stark contrast to the dark rock on which it is built. Seen from this distance it is a staggering thought to imagine that waves actually reach its lantern. No wonder that to countless generations of keepers this place is simply referred to as The Flugga.

Below: Since its automation in 1995, the light flashes continuously over some of the wildest and most inhospitable waters around the British coastline. *(Author)*

11 WOLF ROCK
the curse of the Cornish wreckers

IT WILL BE no surprise to find that for this chapter we return once again to the waters around Cornwall, and to Land's End in particular, where the wrecking communities of these shores reaped a full and bountiful harvest, winter and summer alike. With such a menacing coastline and a marked absence of navigational aids the Cornish plunderers led a rich existence from the numberless vessels lost beneath black cliffs or washed ashore from off the teeth of an ugly reef. If the sea alone could not provide a sufficient crop of heavily laden vessels, then the showing of false beacons from cliff tops made sure a good deal more met with a premature end.

One reef in particular, a cruel benefactor for all who elicited a living by preying on shipwrecks, was devastating in its effects. Its existence has already been remarked upon in an earlier chapter. Found 8 miles south-west from the point of Land's End is the Wolf Rock, a low serrated crag of granites, greenstones and phonolite rising abruptly from 20 fathoms to break the surface in the channel between the mainland and Scilly. Extending for some 175

Above: Now that it is automated, the power for its warning flash comes from solar energy. The ring of solar panels that generate this power can be seen encircling the lantern.
(Steven Winter)

yards in length by 150 yards in width, this reef is completely covered at high spring tides, showing itself only when these waters have receded as a plateau of rock sloping gently to the south-east. It is the focal point for its own peculiar tidal irregularities which swirl around the reef and are experienced at some considerable distance from it. Rising from a deep water channel, the Atlantic waves throw themselves upon its crown with incalculable fury, offering no second chance to the careless sailor caught in the grips of its tidal whirlpool.

The name of this reef appears not to have remained constant. Early sources can be found which record the existence of a 'Gulfe' or 'Gulph' rock in the channel to the west of Mounts Bay, although since the beginning of the 19th century the name of 'Wolf', possibly a corruption of its earlier title, has passed into common use. Legend has it that on the reef was a hollow cavern, into which the waves washed, compressing the air inside and forcing this to escape with a howling cry, eerily mimicking the call of a wolf and ensuring for this reef a name fitting its strange phenomenon. Such a piercing wail was invaluable for giving an early warning of the rock, particularly in still, breathless air when the sound would carry for long distances over the water. This did not please the wreckers, who relied heavily on the reef for their plunderings, and who at one time silenced the Wolf by filling the hollow with boulders transported by boat from the shore. With its voice gone, ships once more fell victim to this hungry crag.

Readers will recall that it was as a consequence of demands by sea traders plying between the English and St George's Channels that Lieutenant Henry Smith was granted a lease from Trinity House in October 1791 to erect a lighthouse on the Wolf Rock. This was subsequently amended on 29th September 1795 to a lighthouse on the Longships reef with a beacon on Wolf Rock, such was the extreme nature of the enterprise on the latter site. Smith was to pay £100 rental to Trinity House for 50 years.

An even earlier proposal of 1750 for a buoy bell to be moored in the vicinity of the reef, *"...in such a manner that it should swing clear of the Rock, carrying a Bell upon it, so as to ring by motion of waves, and to give notice of danger,"* came to nothing, for the very reason (if we are to believe contemporary accounts) that, *"...this jingling scheme (of buoy bells upon the English coasts for alarming us) was not accepted, on a supposition that the fishermen (not approving the musick) would remove the bells when they catched no fish."*

Opposite: White water swirls around Wolf Rock lighthouse. This day is fairly placid; during severe storms waves frequently reach the lantern.
(Author)

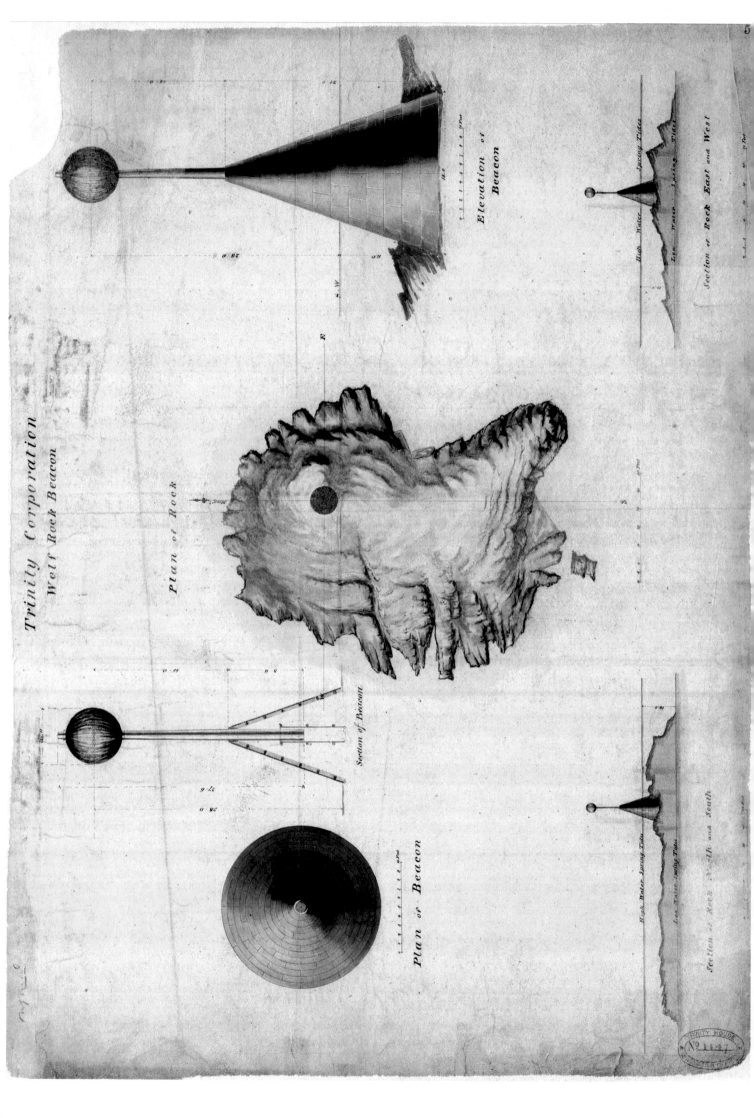

Trinity Corporation
Wolf Rock Beacon

Elevation or Beacon

Section or Rock East and West

Plan of Rock

Section of Beacon

Plan of Beacon

Section of Rock North and South

There was even a scheme, improbable as it may appear, to dynamite the entire reef and thus do away with any further danger once and for all. Henry Smith was a little more philosophical and, four years after he had obtained his original lease, placed upon Wolf Rock a wrought iron mast 20 ft high, 4 ins in diameter, supported by six stays, and topped by a huge replica of a wolf's head cast in copper and with open jaws. It was Smith's intention that his model would recreate the natural howling of the rock as the wind funnelled through the open jaws. Unfortunately, its effectiveness was never really determined with any great certainty as the whole effigy was removed by the first show of might from the sea. It lacked stability at its base and the wolf's head created a top-heavy structure which, although offering little resistance to the waves, was obviously not sturdy enough to resist the colossal weights of water thrown at it during the first storm.

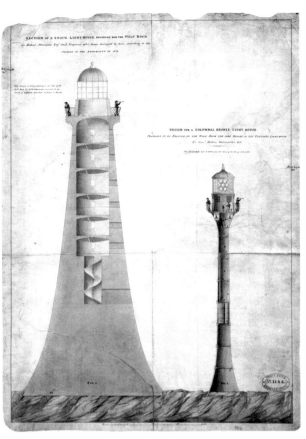

Left: Robert Stevenson, fresh from his triumph on the Bell Rock, visited Wolf Rock in 1823 and prepared these plans for a stone tower on the rock. On the right is a design by Commander Sam Brown R.N. for a "columnal bronze light house...the same height as the Eddystone Lighthouse". Neither were built.
(Trinity House)

The great Robert Stevenson paid several visits to the site on his now famous tours of English lighthouses in 1813, 1818 and 1823. He met with Smith on the first of these occasions and discussed the reef. Smith remained adamant that anything more than a beacon on this rock was, "an impractical work." Stevenson, however, was not totally convinced of this and even went so far as preparing rough plans for the Admiralty in 1823 of a stone tower, requiring something like £150,000 and 15 years for its building. Such a vast sum and time span were considered prohibitive and nothing became of the idea. A Royal Navy commander – Samuel Brown – also produced plans for a hollow column of bronze "the same height as the Eddystone light-house" with a lantern on top of it. He estimated it would cost £16,000 and could be built in 6 months. Neither Stevenson's nor Brown's structures got any further than paper plans.

The next seamarks to actually appear on Wolf Rock were a series of four beacons of varying dimensions, put there by Trinity House from an original design of James Walker but executed by his protégé John Thorburn. The first of these was five years in the building, from 1836 to 1840, although only a total of 302½ hours were actually spent on the Wolf. It appeared as a triangular metal cone, 12 ft across its base, 22 ft high, out of which projected a stout wooden mast of English oak, 12 ins in diameter and surmounted by a 6 ft diameter hollow metal globe. Its total height was in the region of 40 ft, ensuring good visibility above the highest of spring tides. For increased stability the conical metal base was filled with masonry rubble. It cost £11,298-4s-1d.

From completion in July 1840 it lasted only until the November of the same year when the sea removed its upper half. Prompt action was taken to rectify the damage by replacing the oak mast with one of wrought-iron, again topped by a 6 ft diameter metal sphere. Poor weather

Opposite: These are the original 1836 hand-painted plans by James Walker for a triangular beacon to be placed on Wolf Rock before it was decided to build a lighthouse there.
(Trinity House)

during the following summer delayed its completion until August 1842. This mast too suffered much torment from the waves which succeeded in bending it 3 ft from the perpendicular in the direction of the heaviest seas. In October 1844 it was finally snapped, 4 ft from the top of the base.

By July 1845 a third attempt was seen, this time with a 9 ins mast carrying a smaller sphere of only 4 ft diameter. This also was found to have disappeared after a violent storm during May 1848, until it was again duly replaced with another 9 ins mast and an even smaller globe 3 ft across. It is pleasing to report that this, the fourth attempt, lasted until the successful completion of the present lighthouse on Wolf Rock, although there must have been few people indeed who believed it could survive this long after the contemptuous way its predecessors had been treated by the sea.

Left: Even when the sea a short distance from Wolf Rock is comparatively calm, the submarine contours cause huge rollers to break over the rock in a spectacular fashion. This impressive wave was caught by a Wolf Rock keeper from the lantern gallery. *(B. G. Hawkins)*

The first 60 years of the 19th century saw rapid changes and innovative lighthouse engineering techniques. New, isolated sites were being conquered and a maritime journey by nightfall was no longer the nightmare it was previously. The Hanois Rock off Guernsey was built upon, the Bishop Rock and the Smalls were lit by fine masonry towers, all of which led Trinity House in April 1860 to apply to the Board of Trade for finance to be spent on a masonry lighthouse for the Wolf Rock. This was duly approved and James Walker started work on a design for the structure.

Above all, the immediate impression of Walker's plans was the striking similarity between his proposals for the Wolf and the existing beacons on the Bishop and the Smalls. This was hardly surprising as all three sets of plans had originated from the fertile mind of the same engineer. Walker proposed a total height of almost 117 ft with a base diameter of 41 ft 8 ins and an upper diameter of 17 ft. A solid base, except for fresh water tanks, was to extend for 39 ft 4 ins above its base, and the walls varied from 7 ft 9½ ins to 2 ft 3 ins in thickness. He calculated on using 44,506 cu ft of masonry weighing almost 3,297 tons, each block to be dovetailed to its neighbours by having a raised dovetailed band 3 ins high on the top bed and one end joint of each stone. A corresponding dovetailed recess was cut into the bottom bed and end joint of adjoining stones with just sufficient clearance for the raised band to enter it freely for

Below: This early engraving shows the first keepers being put on the Wolf Rock on Christmas Day 1869 from a paddle steamer, appropriately named *Wolf*. It would appear that the landing of the keepers was a difficult task judging by the seas in the illustration.

setting. In addition to the dovetailing, each block of the lowest two courses was bolted to the rock itself, and from the third to the twentieth courses all blocks were again secured by 2 in bolts to the course below.

By the autumn of 1861 James Nicholas Douglass had been appointed resident engineer, coming straight to the site after completing the light on the Smalls. A workyard was acquired at Penzance where a timber jetty was built, and arrangements made so that work could commence the following spring. Before then the boats used at the Smalls would be brought to Penzance to serve again for the same engineer. This flotilla consisted of a 60 hp steam tug plus five barges of 40 tons capacity. Also, a 100-ton schooner was commissioned as a floating barrack to ride at anchor adjacent to the site.

A preliminary survey to decide upon the exact location for the tower was made during 1861, while Douglass was completing his work on the Smalls. Dubious weather had prevented his leaving that site until late June so it was not until 1st July that he could actually land on the Wolf. It turned out that Douglass' first visit to this reef was not what he might have wished for. It was, by all accounts, a most unsatisfactory way of making the acquaintance of a site that was to occupy his attentions for most of the following nine years.

For Douglass, the time spent on the rock was exceedingly short, but he busied himself walking its craggy surface to decide on a site for the base of the tower. A rising tide halted his deliberations but the gathering swell prevented a close approach by the waiting boat. It was clearly a serious situation for the lone engineer as the rising water would soon cover the rapidly diminishing rock. He escaped by tying a life line which he had fortuitously brought

Above: Wolf Rock is probably subjected, more so than any other British lighthouse, to seas of immense proportions. The contours of the rock means that heavy seas driven by a westerly gale can climb the tower to the height of the lantern with alarming regularlity, causing the whole structure to shudder. This dramatic photograph shows the lighthouse in a winter gale of 1971, and before it had a helideck. *(Ministry of Defence, Crown Copyright)*

with him around his waist and leaping into the heaving seas to be rudely hauled into the pitching boat. Although drenched to his skin, Douglass was none the worse for his experience, indeed this method of embarkation was often used in the future by the men when similar situations occurred.

The first working party stepped on to the reef on 17th March 1862 to start the excavation of the foundations. Not only was a circular tower to be set upon this site, but adjoining it, on its north-eastern quadrant, a landing stage was planned. This was born out of necessity as it was only from this direction that the reef could be approached, even then it presented no vertical face to which boats could draw up to land materials. In consequence a substantial landing platform was going to be built at the same time as the tower containing some 14,564 cu ft of material. It was this platform that occupied a considerable proportion of the time in the first season. When the tides did not ebb far enough to admit working on the tower foundations, all the men's energies were directed towards the platform, and so rapid was the progress here that the platform was half complete before a single block had been laid in the foundations of the tower.

The work in these early stages involved considerable danger and called for a high degree of dedication from the men. Being low and flat in profile, breakers frequently swept across the reef carrying men and tools before them. To combat such interruptions, stout iron stanchions were driven deep into the heart of the reef at regular intervals, attached to which were safety ropes so that the gangs of men were never very far away from a rope end. An experienced man was positioned on a high point of the reef to act as 'crow' or look-out, whose job it was to give advance warning of waves likely to sweep the rock. When such warnings were given the men, all of whom were ordered to wear cork life jackets, would grab a life line and bend, head down towards the direction of the breaker while it washed over them. If their tools had not been removed by the force of water they could continue with their labours, although picks and hammers weighing in excess of 20 lb had been lost forever by just

such events. It was slow, erratic and exhausting work for the gangs of men. Work stopped on 29th September with a total of 83 hours gained on the rock during 22 landings. Working every minute the sea allowed, the landing stage rose from the reef and good progress had been achieved with the base excavations.

Shortly after the desertion of the reef for the winter James Walker, chief engineer to Trinity House, died in October 1862. James Nicholas Douglass was appointed to his position, while his brother, William Douglass, succeeded him as resident engineer on the Wolf Rock.

The following year 206½ hours of work were managed in 39 landings between 20th February and 24th October 1863. The landing stage was half complete, so too was the circular trough for the tower. In the workyard at Penzance the sixth course of masonry was cut, dressed and ready to be laid. In 1864 things started to move a little more rapidly. Work commenced on 9th April and by 6th August the first stone of the tower was in place. The landing platform was nearly complete and on to this an iron derrick, 10 ins in diameter, was secured to enable the blocks for the tower to be landed with comparative ease. Such blocks were carried in barges to the site and lifted by derrick from their holds. Elm rollers facilitated their movement within the hold to a position beneath the hatch. Using this method, stones of great weight were landed when the boat was rising and falling as much as 12 ft. At the close of the season on 5th November, 37 stones of the second course (the first entire course), were in position and the tenth course was ready in the workyard. To get this far had taken 42 landings and 267 hours of work.

Left: As the Trinity House helicopter approaches, the keepers wait on a gallery below the helideck ready to emerge through one of the yellow trapdoors. The netting that surrounds the deck is to catch anything that is light enough to be caught by the wind and prevent its permanent loss in the sea. *(Author)*

The season of 1865 was an unusually long one, extending from 11th April to 17th December, during which time the fourth course was completed and 34 stones of the fifth laid. This was in fact the second time the 34 blocks had been set in position, the original stones had been swept away during a storm on 24th and 25th November. It was later presumed that this damage was not entirely the work of the sea but was due in part to floating wreckage as marks caused by the cutting of a chain along the edges of masonry were evident when the weather had moderated sufficiently for an inspection to be made. Soon after the storm it was ascertained that the 1,544 tons *Star of England*, en route from Calcutta to London, was in serious difficulties 100 yards to the weather side of the reef and only avoided striking it by cutting away her mizzen mast with all its gear. It was assumed that this had done the damage to the masonry on the rock. Had it not been for this unfortunate setback the fifth course would have been complete in this season which had seen 41 landings and 250 hours of work The eighteenth course lay ready and waiting at Penzance.

Only the remains of the fifth course plus four more complete layers and ten blocks of the tenth course were set during 1866. It took from 5th March to 13th October to do this in 224½ hours spread over 31 landings. This was also the year when a sturdy iron crane, 16 ins in diameter with metal 1¼ ins thick, was installed in the centre of the tower to ease the task of setting the blocks as the cylinder of masonry rose. It stood 20 ft clear of the stones, 23 ft above the high water of spring tides, but during that winter it was snapped off flush with the working surface.

Landings in 1867 were delayed by the weather until 6th May from when it was possible to work on 40 occasions up until 5th November. Progress was beginning to accelerate and several milestones were passed. On 13th June, with a calm sea, all hands were able to remain on the rock for the first time during a high tide as they worked on the eleventh course. Also in this season, the sixteenth course marked the completion of the solid base. Work speeded ahead to complete the twenty-third course and part of the twenty-fourth which was at the level of the entrance room.

Right: An assortment of gulls rest on the various steps and surfaces of the Wolf Rock landing stage.
(Steven Winter)

A further 23 courses appeared in 1868 between 31st March and 14th October, the result of 30 landings and 276½ hours of toil. For the first time on a tidal site a steam winch was brought into operation from 17th June which speeded progress even further. Just 2½ minutes were now taken to haul blocks from the landing stage to the top of the work compared with almost 15 minutes when raised manually. As the final block of course 47 was laid on its bed of mortar, the dressing of the last piece of masonry in the yard was also completed.

It took only from 16th March to 19th July to complete the masonry in 1869. The final block was laid by Sir Frederick Arrow, Deputy Master of Trinity House. In all eight working seasons, 266 landings had been effected, occupying the men for a total of 1,809½ hours. The time to the end of the season was spent installing the lantern, optical apparatus, doors, windows and internal fitments.

In order to give the Wolf lighthouse a distinctive character a revolving dioptric light of the first order was installed, showing alternate flashes of red and white at half-minute intervals. The colour was obtained by placing ruby glass panels in front of the prisms but revolving with the apparatus. The lantern was one of Douglass' famous helically framed versions which was exhibited at the Paris Exhibition of 1867 before being fitted into the tower. The optics came from the foundry of the famous Chance Brothers and formed, together with the lantern, what Douglass considered to be, "…the most perfect for the purpose that has yet been constructed."

The doors, windows and storm shutters were, as was normal practice, fashioned out of gun metal, so too were the internal staircases to reduce the risk of fire. Perhaps Douglass had recalled the flaming inferno which was once Rudyerd's tower on the Eddystone and was wisely taking all precautions possible to prevent the occurrence of a similar situation here. Even though the tower was built of stone, the cheaper wooden stairs would enable any fire to spread rapidly between floors with disastrous consequences for the keepers who might find themselves trapped on one floor with no method of escape. A huge 5 cwt fog bell was suspended from the lantern gallery and struck three times in quick succession every 15 seconds by two hammers worked from the machinery of the illuminating apparatus.

The finishing touches were supervised by Mr Michael Beazeley, the assistant engineer, as Douglass had left to work on a similar site off the coast of Ceylon. The weather around Cornwall began to deteriorate; on 11th September, during a violent westerly gale, large quantities of water were driven over the top of the tower. The men working inside stated the shock waves of each blow could clearly be felt but scarcely any tremor of the structure was perceptible. By 2nd December the colza oil and stores were taken into the tower, followed on Christmas Day by the keepers who had a difficult journey to take up their posts. James

Douglass made the journey with them but delayed his departure because of his enthusiasm to stay and supervise matters. In order to reach his boat Douglass had again to be dragged through the surf on the end of a line, a vivid reminder of his very first visit to this wild site eight years previously.

From 1st January 1870 the red and white flash every 30 seconds radiated from the top of Douglass' tower. The total cost of the undertaking, including lantern, illuminating apparatus, workyard, vessels and wages for the 70 men was calculated at £62,726 – a sum which bears favourable comparison with other lighthouses on similar sites.

The number of vessels lost in this region fell dramatically, although mention should be made here of the unfortunate loss the 698-ton barque *Manitobah* as a direct result of the lighthouse being brought into service. On 1st February 1872 the captain mistook the new Wolf light for St Agnes, Scilly, as he failed to see the bright red flashes of the former. His vessel struck the notorious Bucks Rocks off the Cornish coast with the loss of all 13 crew members.

The Wolf Rock lighthouse also features in another dramatic incident early in 1948 when foul weather had prevented the scheduled relief of keepers who were now growing noticeably short of provisions. On 7th February Trinity House arranged for a helicopter to drop supplies, with the Pendeen lifeboat standing off in case of difficulties. Despite rough seas and a 40 mph wind, the first supplies delivered by helicopter to any British rock station were successfully landed. The Wolf was also the first lighthouse to have its keepers relieved by helicopters when the new helideck was constructed above its lantern in the early 1970s. This experimental platform was the forerunner of the models now crowning other Trinity House lights and first came into service on 3rd November 1973 when the keepers were relieved by this method, although the helicopter did not actually set down on the pad but winched the keepers into it.

Left: The crevices and gulleys of Wolf Rock are subjected to really ferocious waves, tides and currents where only the most tenacious seaweeds can survive. At low tide the rock takes on a noticeable pink hue and provides a shelter for wildlife.
(Steven Winter)

I have remarked in the previous chapter, but it is again pertinent here, that while such progress as the adding of helidecks is desirable, and the use of such structures will obviously spread to further exposed lights around the globe, it is with some regret that the glittering beauty of the lanterns crowning such magnificent towers will never be prominent as they once were. The mortar board headgear now worn by these sentinels does nothing to flatter the graceful beauty of their tapering forms, evermore cast into gloom beneath a steel shroud.

Before the historic helicopter relief, the original lantern apparatus was removed and replaced by a modern electric 1,000-watt bulb, lit by generators installed in one of the lower rooms. The character of the flash remained as originally installed but the range increased to almost 23 miles in clear weather. The fog bell was also superseded by a modern diaphone horn, sounding for 4 seconds in every 30.

Automation took place at the same time as at Longships to make best use of resources in the Penzance area. A LANBY was positioned a short distance from the rock while the main navigation light was extinguished. The old equipment was withdrawn and replaced by new generators, long running filament lamps, standby light and electric fog signals and detector – almost an identical program to that at Longships. Automation was finally accomplished in March 1988. In May 2003 the light at Wolf Rock was extinguished again – not through any breakdown or malfunction, but for the process of conversion to solar power. A lightvessel

Above: The swirling currents around Wolf Rock are seen to good effect in this view from the Trinity House helicopter. *(Author)*

named *Wolf Rock* was towed into position 1¼ miles off the rock while the rings of solar panels were attached to the helideck supports. By the end of September it was done and the lighthouse once more sent out its reassuring flash. Wolf Rock is now monitored from the Trinity House depot at Harwich.

Its automation means that there will no longer be anyone inside Wolf Rock to experience what is regarded as a unique phenomenon at this particular rock light – something that probably never occurred at any other tower. It was passed to me by a former keeper, albeit it via an intermediary, who felt it was of sufficient interest to merit recording in this volume. It concerns an effect felt inside Wolf Rock in times of storm. At sea the speed of the wind isn't moderated by the landscape – because there isn't any. There are no 'gusts' of wind as such – just a solid continuous shrieking blast. From time to time in a full gale the Wolf was often completely enveloped in rising spray, and when this happened there was, until the water fell away, a complete cessation of wind-roar, a moment of almost total calm in the midst of a roaring tempest. Whether this unique feature of the Wolf was because it lacked the cylindrical base of Eddystone and Bishop Rock we shall never know, because it's very unlikely to be heard by anyone ever again.

The Wolf was the final lighthouse in the group of famous beacons which warn of specific dangers along this particular morbid stretch of our coastline. Following as it did the Eddystone, Longships and Bishop, these once feared rocks had all been lit by towering beacons, each one a splendour to behold and each one a saviour to successive generations of mariners.

The final word in this chapter must lie with a member of the Institution of Civil Engineers who, after listening to James Douglass delivering his paper on the construction of the lighthouse was forced to remark that he knew, "…of no site or position for a lighthouse which required more skill, or involved more difficulties or dangers, than was the case with the Wolf Rock structure." Furthermore, he regarded its erection, "…as one of the greatest boons to the commerce of the country. It would be the means of saving thousands of pounds worth of property in a year, beside human lives. Instead of that spot being shunned, it would be now anxiously sought for as one of the best landmarks that could be obtained." There could be no finer tribute than that to any lighthouse.

Opposite: Since automation Wolf Rock still flashes its warning, but for 24 hours a day. In the distance we can see a ship giving the rock a wide berth. *(Author)*

12 DUBH ARTACH

Guardian of the Hebrides

SITUATED in the same stretch of ocean as the magnificent Skerryvore, not 20 miles distant from it, rises another of the great Scottish rock lighthouses. Again, it is a further product of the Stevenson family, and accordingly exhibits the high standards of functional beauty associated with these men. David and Thomas Stevenson, the second and third sons of Robert, were responsible for its construction while in their capacity as joint engineers to the Commissioners of Northern Lighthouses. They fought nature with calculation and built upon a seemingly impossible site in mid-ocean to continue the tradition of this remarkable family, overcoming impossible odds at a situation equally as exposed as any of their predecessors had worked on.

One of the principal sea routes in Hebridean waters is the Firth of Lorne – a wide tapering channel giving access to the once bustling ports of Oban and Fort William. Its mouth is some 60 miles across and marked by lighthouses built on the Rhinns of Islay in 1825 and Skerryvore in 1844. Between these two beacons is a considerable stretch of unlit water. Normally this would have been of little consequence had it not been for a particularly devastating reef known as the Torrin Rocks, 4½ miles of jumbled granite teeth breaking water some miles off the Ross of Mull. The extent and confused nature of this reef claimed untold numbers of vessels plying between America or the Baltic ports and Oban.

As if this were not danger enough, 9 miles beyond the outermost rock a solitary outlier rises from the waves, a huge whale-back rock of spectacular dimensions – 240 ft long, 130 ft wide, rising 35 ft above normal spring tides. Its massive bulk is composed of a hard, dark basaltic rock, so hard that countless centuries of wave action have been unable to eat into its almost vertical faces. Around the rock there is not one solitary creek or gully into which a boat could take shelter from the enormous waves that wash over this colossus. It has come to be known variously

Left: Dubh Artach as seen from the deck of an NLB tender.
(Eddie Dishon)

as Dubh Artach, Dhub Artach, Dhu Heartach or St John's Rock and was the site chosen by the Stevenson brothers for the lighthouse which would not only warn of the rock itself but give valuable prior warning of the Torrin Rocks and help vessels to find a safe anchorage amongst the islands of the Inner Hebrides. Its name is most likely a corruption of the gaelic *an uibh-hirteach* meaning 'the black one of death', but its anglicised version of Dubh Artach has been generally adopted since 1964.

Robert Louis Stevenson was moved to write of the rock:

An ugly reef is this of the Dhu Heartach; no pleasant assemblage of shelves, and pools, and creeks, about which a child might play for a whole summer without weariness, like the Bell Rock or Skerryvore, but one oval nodule of black trap, sparsely bedabbled with an inconspicuous fungus, and alive in every crevice with a dingy insect between a slater and a bug. No other life was there but of sea-birds, and of the sea itself, that here ran like a mill-race, and growled about the outer reef for ever, and ever and again, in the calmest weather, roared and spouted on the rock itself.

The Commissioners of Northern Lighthouses had long been aware of the rock but considered it an impossible site to build upon. In June 1864 Thomas Stevenson, the father of Robert Louis, travelled out to the reef and was fortunate in actually being able to land upon it, where he emphasised further the need for some kind of warning. The urgency of the

Opposite: Dubh Artach lighthouse was built on a solitary whale-back hump of volcanic rock between 1867–72 by David and Thomas Stevenson.
(Air Images Ltd)

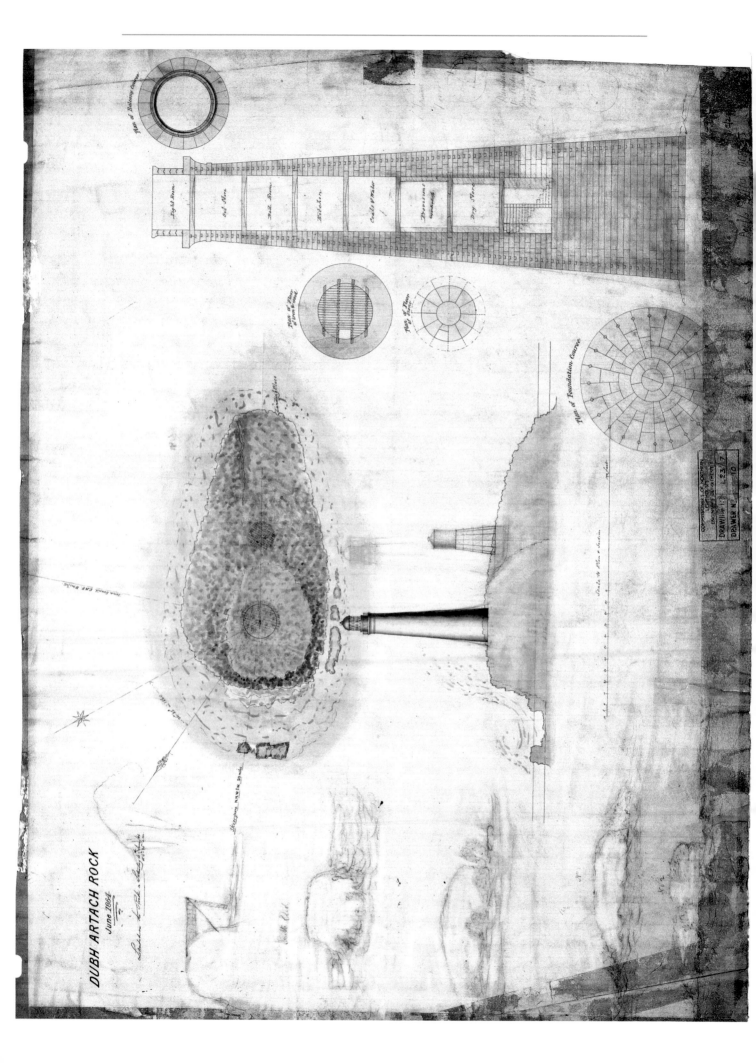

DUBH ARTACH ROCK
June 1864

Plan of Balcony Course

Light Room
Oil Store
Bed Room
Kitchen
Coals & Water
Provisions Cisterns
Dry Store

Plan of Floor of Iron Gallery

Plan of Floor of Store

Plan of Foundation Course

Scale to Plan & Section

undertaking was vividly demonstrated during December 1865 and January 1866 when storms of seismic fury tore along the west coast of Scotland, wrecking 24 vessels in this vicinity on the night of 30th and 31st December 1865.

As a result of these losses greater pressure was brought to bear on the Commissioners who, after examining the records of wrecks, finally obtained the statutory sanction from the Board of Trade on 20th October 1866 when the Stevensons began work on their plans. During preliminary surveys the engineers had seen the exceptional violence with which the waves threw themselves against the rock and the awesome distances they climbed into the air. Its position, although exposed, was no more so than Skerryvore or the Bishop Rock, yet this site was subjected to seas of spectacular proportions. In his account of the building of the lighthouse David Stevenson remarked that, "Owing to the peculiar formation of the Rock the impulsive force of the waves is led up to a much higher level than in any other lighthouse tower with which we are acquainted. It is believed that at no lighthouse tower hitherto constructed have such remarkable proofs of the violence of the sea at high levels been observed."

For the reason behind this peculiar phenomenon we must look at the submarine contours in the vicinity – as determined by an Admiralty survey. These show clearly that Dubh Artach lies at the head of a submarine valley, 80 miles in length and stretching away from the rock towards the open Atlantic. Stevenson thought, "…it is impossible to doubt that a funnel-shaped deep track, receiving directly the seas and currents of the Atlantic Ocean, must have some effect in concentrating the waves on the lighthouse rock at the head of this submerged valley, and may therefore account for the seemingly abnormal seas to which the tower is subjected…" Knowing this, the Stevensons realised that a tower of the utmost strength was required on Dubh Artach, able to withstand pressure of 3 tons to the square foot, if it was not to be swept away by the first gale of the winter.

The final form produced by the engineers was for a tower based on a parabolic frustum, 107 ft high, 37 ft in diameter at the foundations and 16 ft across the upper course of masonry. Within its graceful curve six rooms were encased, entrance being gained through a door 32 ft above its foundations. At this stage they estimated it would cost £56,900.

The Stevensons received authority to start their work on 11th March 1867, leaving precious little of that season in which real progress could be made. However, a shore station on the Isle of Earraid was chosen in preference to Tinkershole on Iona's eastern coast as granite was found there in abundance and there was sufficient room for the erection of permanent buildings for the masons and seamen. A wharf was constructed for the servicing of the boats, no harbour being required as the Sound of Earraid was suitably sheltered. Although it was 14 miles distant from Dubh Artach, it was considered the most appropriate site for continuous communication with the rock.

The first season also saw the first steps taken in the erection of a barrack on the reef in which the men could be housed when work was in progress. This, of course, meant that progress was possible even though it may have been too rough to land from a boat. Between 25th June and 3rd September, 27 landings were made from the chartered steam tug *Powerful* out of Leith, during which time excavation of the tower's foundations was commenced and the first tiers of circular wrought iron supports for the barrack were set in position. Over the winter, a paddle steamer *Dhu Heartach* was built to tow the lighters loaded with stone out to the rock, as well as the construction of cranes and winches, etc, at the shore station.

The new steamer arrived at Earraid on 14th April 1868 in time for an expected early start to the proceedings. Normally, easterly winds prevailed in this vicinity during the spring, making landing on Dubh Artach somewhat less hazardous. In this particular year, however,

Opposite: A very colourful plan for Dubh Artach lighthouse that has been annotated by someone that they "landed on the rock on Saturday 4th June /64 at 4h-30m a.m." It shows where the proposed light was to be built and the iron barrack alongside it. However, the profile of the tower in these plans – with straight tapering sides – was not the tower that was eventually built. (*Northern Lighthouse Board*)

the strong westerlies continued until the end of June, severely dashing all hope of an early start. Only two landings were possible in these first two months, the first on 4th May allowed a handful of men to leap on to the rock for an hour and a half – time enough for them to inspect the previous season's efforts and to note that only one section of ironwork from the barrack support was missing, 30 ft above the high spring tides. The visit of six men on May 18 was equally brief, a rising sea causing their premature departure. Not until 29th June did the work start in earnest, and from then until 28th September a total of 38 days' work was possible, which resulted in the completion of the barrack and three-quarters of the excavations.

The principle of the barrack was exactly the same as the structures seen on the Bell Rock and Skerryvore – a refuge for the men during the season removing the need for time-consuming journeys to and from the rock, and doing away with the uncertainty of landing. Whereas the barrack at the two former sites were of wooden construction, the Stevenson brothers had wisely decided on iron for Dubh Artach in view of the exceptional violence of the waves at this site. The malleable iron legs raised the covered portion 35 ft above the rock. They were 4 ins square in section and sunk 3 ft into the rock. On top of these the living quarters, composed of riveted plates ¼ in thick, were 18 ft 9 ins high and 16 ft in diameter. The inside was divided into two storeys.

By August it was complete and occupied by one of Stevenson's foremen, Alan Brebner, together with 13 men. In order that they may press ahead with all speed, it was decided to use the barrack during this season, even though it remained untested by heavy seas. On the night of 20th August huge seas rose without warning and hurled themselves against the metal drum with menace. On many occasions the waves rose higher than the top of the barracks, itself 77 ft above high water, to crash down on the roof, deafening the inhabitants and excluding all light from the windows for seconds at a time. The sheer power of the waves as they swept across the back of Dubh Artach was responsible for bursting open a hatch in the floor of the lower tier, held closed by a substantial iron bolt, but still 55 ft above the sea level, flooding this compartment before draining out again and taking with it vital provisions. It was five days of hell before the seas were subdued enough for a boat to approach and rescue the frightened inhabitants of the barrack which, although tested to the extreme, still stood firm. The following graphic description of life in the barrack has come from the gifted pen of Robert Louis Stevenson in his Memories and Portraits.

When the dark nights fell, and the neighbour lights of Skerryvore and Rhuval were quenched in fog, and the men sat high up prisoned in their iron drum, that then resounded with the lashings of the sprays. Fear sat with them in the sea-beleaguered dwelling; and the colour changed in anxious faces when some greater billow struck the barrack, and its pillars quivered and sprang under the blow. It was then that the foreman builder, Mr Goodwillie, whom I see before me in his rock-habit of indecipherable rags, would get his fiddle down and strike up human minstrelsy amid the music of the storm.

Meanwhile, at Earraid, the work progressed rapidly, buildings rose, the wharf was improved and a large proportion of the lowest courses of the stone were dressed and ready. The year of 1869 saw just 60 landings on the rock between 25th March and 28th October. Stevenson remarked that, "...our anticipation as to the difficulties of making landings at Dhuheartach have been most fully realised..." The workmen occupied the barrack between 26th April and 3rd September, giving a working season of 113 days, Sundays excluded. During this span, the foundations were completed, the first stone being set on 24th June, and the first course finished six days later. These were good omens – but they were not to last.

A severe gale on 8th and 9th July drove the men off the rock for six days and removed 14 stones of the third course, each of 2 tons weight, set in Portland cement and held in position by metal joggles. All this took place 35 ft above high water. Eleven of these blocks were never seen again – the waves had carried them completely off the rock. By way of minor damage the landing cranes were completely destroyed. Stevenson wrote that, "It is a remarkable fact that the level above high water – 35 feet – from which these blocks were removed by a summer gale is the same as that of the glass panes in the lantern of Winstanleys first lighthouse on Eddystone, which nevertheless stood successfully through a whole winter's storms."

The devastation of the summer gales of 1868 and 1869 caused the engineers to have second thoughts about their original plans. Assuming that the winter's storms would be of equal, if not greater violence, they wisely decided to increase the height of the solid base by 11½ ft, raising it from 52 ft 10 ins above high water to a remarkable 64 ft 4 ins above high tides, giving it a weight of 1,840 tons. When compared with other rock towers, the height of Dubh Artach's solid stump is in a class of its own; Smeaton's Eddystone rises only 10 ft 3 ins, Bell Rock 14 ft, while even the Bishop Rock's solid trunk is only 23 ft high. The nearest rival to Dubh Artach also happens to be its nearest neighbour – Skerryvore, which can only manage 30 ft 6 ins.

After the seas had subsided, work recommenced on 16th July. By 30th August the fifth course was in place and the remainder of the season was spent securing existing blocks and the removal of cranes. On shore five dwelling houses were ready and 17 courses of stone lay waiting. The scene at Earraid is again admirably described by Robert Louis Stevenson:

Top: Before despatch to the reef all the stones were fitted together on a 'stage' at Earraid to check for accuracy.
(Northern Lighthouse Board)

Above: The solid base is well under way in 1870. The stones were moved from the jetty to the tower on an inclined tramway. The trucks were hauled alone the slope by the steam engine on the right of the picture.
(Northern Lighthouse Board)

There was now a pier of stone, there were rows of sheds, railways, travelling cranes, a street of cottages, an iron house for the resident engineer, wooden bothies for the men, a stage where the courses of the tower were put together experimentally, and behind the settlement a great gash in the hillside where granite was quarried. In the bay, the steamer lay at her moorings. All day long there hung about the place the music of clinking tools, and even in the dead of night, the watchman carried his lantern to and fro in the dark settlement and could light the pipe of any midnight muser.

The season of 1870 started late. The first landing of men was made on 14th April, but it was not until 6th June that the lighters could approach with a cargo of stones. Each block was shaped, dressed and fitted with its neighbours at Earraid. Sixteen tons of blocks were loaded into each barge which was then towed, usually in a pair, by the *Dhu Heartach* for the three hour journey to the rock. They were unloaded by steam cranes and moved around the rock on inclined tramways, again with the help of steam engines. Each block was set in its final position in Portland cement, and held in position by either stone or cast iron joggles, all this work being under the superintendence of Mr Alan Brebner.

After the slow start, conditions quickly became more favourable. The thirty-first course was complete on 24th August, and the rock abandoned six days later, almost five months and 62 landings since the season started. The engineers also deemed this time prudent to order the lantern casing and optical equipment.

As the structure climbed further out of reach of the waves, 1871 passed without undue trouble. A total of 57 landings were made between 1st May and 10th October. All dwellings at Earraid were finished, but more importantly, so were all 77 courses of masonry, the upper course reaching 101 ft above the foundations, giving a total weight of stone of 3,115 tons.

It was with light hearts that the men must have stepped on to the rock on 23rd April 1872. All that remained before their task was complete was the assembly of the lantern, the installation of the optical apparatus and its mechanism, plus the internal fixtures and fittings. Although a relatively uncomplicated task, it took most of the year. During this period a party of Commissioners visited the tower on 24th July 1872, landed "without difficulty" to report that they, "were extremely well pleased" with what they found. When all was finished, the centre of the lantern was about 145 ft above the sea. The original equipment showed a fixed white light visible for 18 miles except for an arc of 7 degrees which showed red. It was lit for the first time on 1st November 1872. The following day James Ewing, one of the

Left: What looks like a rather 'wet' relief day at Dubh Artach. The relief is taking place to the right of the tower amongst a worrying amount of white water. The remains of the barrack legs are still visible on the left. *(Northern Lighthouse Board)*

keepers, wrote to the secretary of the Lighthouse Board, "I beg to inform you that the light was exhibited on the night of the 1st to conform to your orders. I am glad to say that we carry a magnificent flame, which eventually must eclipse all the lights on the west coast." The new fog bell machinery was of novel construction, producing ten seconds of rapid strokes every 30 seconds to distinguish it from the single stroke of Skerryvore every minute.

Shortly after the lighthouse became fully operational, the Stevenson brothers were to bear witness to yet another example of the might of the waves that assail Dubh Artach. The tower had not yet given a week's service when on the night of 5th November 1872 another winter gale struck the west coast of Scotland. At Dubh Artach the huge waves spent their force against its new masonry pillar with sufficient force to wrench the copper lightning conductor, a 1½ ins by 1 in band of solid metal, from the channel in which it ran so that its outer surface was flush with the stonework. The screws holding it in position were torn out of their sockets. Even more disturbing was the fact that this conductor was on the lee side of the tower and bent at a level with the kitchen window, 92 ft above high water. In spite of these assaults, the tower survived the winter, although during December James Ewing reported that the light produced in Dubh Artach was of such brilliance that it was affecting the eyes of his assistant keepers.

During 1873 the barrack which had served so well during the construction was finally dismantled. The lower portions of the legs were left in position as the expense of removing them would have exceeded their scrap value. This circular iron framework was, until quite recently still visible on the rock, over a century since it was first placed there – the rusty latticework of metal providing a perfect attachment for rocket lines, etc.

When the final accounts for the undertaking were presented, it was seen that Stevenson's earlier estimate of almost £57,000 was extremely optimistic. The bill for just the tower, lantern, optical equipment and fog bell came to £65,784 , while a further £10,300 was spent on the nine shore dwellings and wharf at Earraid, giving a final expense of £76,084. David Stevenson, in order to qualify this seemingly large amount, commented that:

When it is considered that Dhu Heartach Rock is of small area, surrounded on all sides by deep water, and exposed to the swell of the Atlantic Ocean, that it is nearly 16 miles from the shore, and could only be worked on for about

Above: The Stevenson brothers experienced regular examples of the violence of the waves at Dubh Artach caused, it is thought, by the rock being at the head of a submarine valley. The amount of white water in this 1973 view is certainly not unusual.
(Northern Lighthouse Board)

two months and a half in the year, we believe that the amount of work executed will bear favourable comparison with other undertakings in less exposed situations; while the fact that the whole was brought to a successful close without the loss of life or damage to property bears ample testimony to the skill which distinguished the personal conduct of the rock work and shipping of this difficult undertaking.

All the workmen employed to build the lighthouse were paid on a day-to-day basis, it being uncertain exactly how often work would be possible on the rock. This situation led to a novel and interesting arrangement being made for the transport of wages out to the men. During the early months of the operations a man would leave Earraid to collect the money from

Right: The automation of Dubh Artach is in full swing. Scaffolding has been erected up to the height of the entrance door, and a temporary light has been fixed to the top of the lantern dome. *(Eddie Dishon)*

a bank in Oban. This involved a return journey of some 40 miles across the rather uncertain roads of Mull where he would catch an open post boat into Oban. Return with the money, in the form of silver coins, required the messenger to hire some form of transport back across Mull, such was the great weight of the coinage. All this occupied one week in every month. An arrangement was subsequently entered into with The Royal Bank of Scotland in Glasgow, which for the sake of secrecy was referred to as 'The Box'. This was indeed a wooden box containing between £600 and £1,000 which was fastened to the deck of the *Dunvegan Castle* with a stout chain and padlock by its clerk. The bank in Glasgow and the cashier at Earraid were the only two key-holders. This arrangement continued until November 1871, by which time £26,500 had been conveyed without the loss of a single penny.

Dubh Artach today, like most other stations of the first order, has seen many changes. The most obvious external difference is its acquisition of a broad red stripe around its midriff to help with identification. This was added in 1890 in response to the wishes of local mariners who thought that confusion between Dubh Artach and Skerryvore during daylight was a possibility. David Stevenson wrote that:

In the case however of Dhu Heartach and Skerryvore which are both on low lying rocks, both granite towers and are only 21 miles apart, there may be some difficulties with sailors not thoroughly familiar with their appearance distinguishing the one from the other...In these circumstances I should recommend that Dhu Heartach should have a dark red band 30 ft in height painted round it in three coats of paint, the grey granite above and below this will not require painting.

I regret to recommend anything which would in the least interfere with the architectural effect of Dhu Heartach tower which is to my mind one of the best architecturally in the service suiting as it does so well the rock on which it stands. But while I do not think that the red band which I recommend will seriously interfere with or destroy the appearance of the tower, I am in any case, strongly of the opinion that in lighthouse matters appearance must, if necessary, give place to utility.

The lantern and its equipment have been modernised and now a group flash of two every 30 seconds is shown. More significant is the fact that Dubh Artach was converted to automatic operation on 7th October 1971 – one of the earliest rock lights to be converted. Further conversion to solar power in the 1990s meant the acetylene burner lamp could be removed. The new lamp is monitored from the NLB Headquarters in Edinburgh, while routine maintenance is effected by helicopter. The difficulties of access onto Dubh Artach rock, a major consideration in the automation of this station, were graphically described by one Scott Dalgleish writing in *The Times* of September 1881 after a visit to the light:

As we approached the lighthouse, we could see the keepers at work on the lee-side getting ready the derrick by which we were to be landed. Up to this time it had been doubtful – so said both the captain and the mate – whether the sea would be calm enough; but when we got under the lee of the island, we were assured that landing was quite practicable, though to our unaccustomed eyes the lumpy sea breaking in white waves did not promise much comfort. When we were within a furlong of the island the little steamer dropped anchor. The long-boat was launched, manned by four sailors, and steered by Captain Irvine himself, and in this we were rowed to the south-eastern end of the rock. The process of landing is interesting, though when experienced for the first time it must appear rather sensational to those of weak nerves. The boat is not allowed to touch the rock. It is anchored by a long line stretching seawards, so as to allow its stern to swing within 10 or 12 ft of the rock. The boat is kept in position under the derrick by two stern lines attached to the rock. The derrick consists simply of a spar, which is rigged up by the lighthouse keepers as often as it is required, and from which a stout rope working in a double pulley is suspended. When the boat is in position, the rope,

Right: The base of Dubh Artach is solid up to the level of the entrance door – a huge height compared to other rock lighthouses. The rectangular extension to the lantern gallery railings gives room for a bank of solar panels behind them.
(Northern Lighthouse Board)

which has a loop at the end of it, is dropped into the stern. You put one foot into the loop, hold tightly with both hands below the block, and are first hoisted into the air and then pulled downwards to the rock. There you are clasped in the strong arms of one of the keepers; and before you are released from the friendly grip, you are reassured by a kindly voice bidding you 'Welcome to Dhu-heartach!'

The view of the lighthouse from the rock is totally different from that obtained of it from the sea at a distance of a mile. In the latter view you take in at once all its proportions; and while the frustum which forms its base is, perhaps too broad to give the notion of elegance, the upper part stands out from the sky as a slender and graceful shaft. When you see the lighthouse from the rock on which it stands you lose the general outline; you see only the massive details, and the one idea impressed on the mind is that of tremendous strength. This idea is intensified when you walk round the granite cone, which seems as immovable as the rock in which it is securely embedded.

So the third great Scottish rock tower was once again a product of two brothers from the Stevenson family, whose ability at executing a seemingly impossible project had only served to underline the brilliance of their designs. Even away from the field of civil engineering the Stevensons managed to make their mark in whatever profession they became involved. Thomas's son was no less a person than the great Robert Louis Stevenson who, much to his father's dismay, did not wish to make his career in the family business, but went on to excel with his own particular style of narrative and verse.

As an illustration of their confidence at whatever project they found themselves engaged upon, David and Thomas, while in the middle of Dubh Artach's construction actually started work on another rock tower! It was all that many engineers could achieve to produce one tower, yet the two brothers actually started work on a second tower on a half submerged reef while their present project was only half complete. It is not difficult to see how such men could instil the greatest confidence in their workforce by taking decisions such as this. To continue the exploits of the Stevenson brothers we must move further south and into another chapter.

Opposite: A dramatic view of Dubh Artach lighthouse from a passing helicopter. The circular concrete helipad was constructed on the site of the iron barracks that was used in the construction of the light. *(Colin McPherson)*

13 CHICKEN ROCK

A Manx crag

KEEN ORNITHOLOGISTS will doubtless be aware of the sea bird which breeds along our western shores known as the storm petrel (*Hydrobates pelagicus*). A close relative of the rather more well-known fulmar and shearwater, it's a black, swallow-like bird with a patch of white in front of its tail. They nest in burrows or rock clefts and their diet is normally small surface animals but they are also known to follow, and alight upon, ships to pick up scraps.

The relevance of this piece of natural history in a book concerning lighthouses may not be immediately obvious, but when one learns that the other common name for the storm petrel is Mother Carey's Chicken, after its characteristic perching behaviour, its inclusion in this particular chapter will perhaps be a little clearer. Along its extensive breeding area a favourite perch for this bird were a couple of low, rocky crags which broke surface a mile to the south of the Calf of Man – a small island off the southern tip of the Isle of Man. So popular was this spot with the storm petrel it virtually christened itself Chicken Rock.

At low water over 4,000 sq. ft of bare rock appeared as two rugged islets connected by a low isthmus, all of which was covered to a depth of 5 to 6 ft at high tide. A reef which disappeared regularly with every tide was menace enough, but there were two further evils which beset this malevolent outlier. Firstly, it rose from comparatively deep water into the middle of a rapid tidal race which swept around this southern extremity of the Isle of Man, a tideway which was capable of getting hold of a vessel and dragging its hull across the reef. Secondly, this vicinity is notorious for the thick, blanketing sea-fogs that descend with alarming frequency to blot the reassuring landforms from view.

While the Irish Sea was not normally regarded as an area fraught with as many maritime hazards as, say, the waters of the Hebrides or the coastline of Cornwall, the frequency with which ships bound to or from Liverpool ended their days on the Chicken Rock drew the attention of the Commissioners of Northern Lighthouses (whose jurisdiction had been extended in 1815 to cover the Isle of Man) from Scottish waters whose lighting was occupying most of their time and funds.

Chicken Rock was considered by those who knew these waters intimately to be equally as devastating in its effects as the Bell Rock or Skerryvore, for not only was it visible solely at low tide, it rose sharply from the ocean floor into the middle of a fierce tidal rip in an area which was frequently blanketed by fogs. With such a combination of lethal characteristics, the site quickly became an obvious candidate for some kind of warning apparatus.

Shortly after the Commissioners had inherited the responsibility for Manx lighthouses, an experienced shipmaster was stationed on Calf Island to make detailed observations of the weather. This would no doubt confirm to the Commissioners what local people could have told them had they bothered to enquire, yet out of this seemingly pointless exercise came some good. It was decided to erect two stone lighthouses on the Calf which would warn, not only of the island, but also of Chicken Rock by a rather ingenious system. The two towers were built by Robert Stevenson 560 ft apart and in a direct line with Chicken Rock. One tower was 55 ft high and the other 70 ft high, so that when the two lights were directly above one another, as seen from the sea, the observing vessel was in a direct line with the offending rock. It was a novel attempt to mark the reef but its shortcomings were obvious – it was impossible to give an accurate position for the rock, only the bearing on which it lay could be identified and not its exact position. In heavy seas or poor visibility this could be fatal. Nevertheless, any clue to the whereabouts of Chicken Rock was a vast improvement on the previous inky blackness and the two towers came into service in February 1818.

The sea mists continued to plague this area with such persistence that on many occasions there might just as well not have been the Calf of Man lights as they were completely

Opposite: Almost high water on a windy day at Chicken Rock with the waves sweeping over the reef itself. Now fully automated and powered by solar energy, the light flashes continuously. (*Sue Jones*)

invisible from the sea. Under such conditions the protection afforded was completely inadequate. This led to a resolution being passed by the Mercantile Marine Service Association of Liverpool and forwarded to the Board of Trade on 13th November 1866. It suggested that, "The night light on the Calf of Man, now so often enveloped in fogs and so rendered useless, to be removed to the Chicken Rock which is one mile and a half out and is a rock of considerable size and great danger."

This report was presented to both David and Thomas Stevenson, Engineers to the Commissioners, and also to Trinity House. Both parties were in full agreement with the proposal, Trinity House pointing out in a letter to the Board of Trade on 21st November 1867 that, "...the Calf of Man Lights were not to be depended upon from the well known prevalence of fog on the Calf even when other high lands are clear..." On 6th April in the following year sanction to proceed with the erection of a lighthouse on Chicken Rock was granted by the Board of Trade. David and Thomas Stevenson prepared plans for an appropriate tower, while Dubh Artach, 160 miles further north was only half complete. It was a unique situation, for never before had any lighthouse engineer been involved in building two rock towers at the same time while commuting between the two to maintain and observe progress.

Left: A rare photograph showing the finishing touches being put to the Chicken Rock lighthouse in 1873. The tower is almost complete with many of the blocks that will make up the final courses in various stages of transport to the top of the tower.
(*Northern Lighthouse Board*)

As the site was so close to the Manx coast the brothers wisely decided that a barrack built on the rock or floating nearby would be unnecessary. Instead, a base was started at Port St Mary, a small fishing village 4½ miles from Chicken Rock, where over a hundred masons, blacksmiths and labourers busied themselves with the many details of the project. A tramway was laid between the workyard and pier for the speedy movement of stone. Granite was again the choice of the engineers, quarried at Dalbeattie in Dumfriesshire. It was brought to Port St Mary by sea, cut and shaped from templates, and numbered in the yard, before being loaded into two lighters towed behind the steam tugs *Terrible* or *Dhu Heartach* which had once assisted shipping along the Clyde, for the short journey to the reef.

As can be imagined, a rock which is immersed twice daily gives precious little time to get an appreciable amount of work completed, particularly if heavy seas cause a premature abandonment of the site. With a working season that extended only from April to September, David and Thomas worked doggedly to complete the foundations of their tower during the first season. Thirty-five men toiled and sweated on Chicken Rock for periods which might only last for a couple of hours at a time. Every precious minute was seized in an attempt to complete this crucial work before the winter's gales made all work impossible. It was dangerous, unnerving work; swinging heavy tools and trying to remain upright on a seaweed covered rock, while at the same time watching for the largest breakers that crept in to wash over the rock, was a skill that involved great dexterity and co-ordination. Each man was issued with a cork life jacket which it was compulsory to wear as a safeguard should the unfortunate occur. Happily, throughout the whole five seasons of the project, no lives were lost. Thanks to the enthusiasm instilled in the men by David and Thomas, the circular pit of the foundations was eventually completed shortly before work stopped for the winter of 1869.

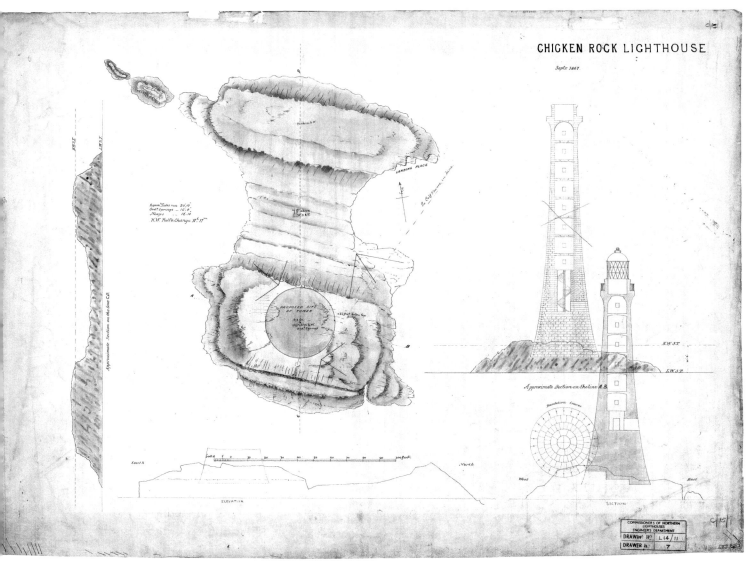

Shaping and drilling of the blocks continued through the winter when the workyard at Port St Mary echoed to the sound of metal striking metal and crumbling stone. During 1870 nine complete courses of the solid base were laid, using just about every possible method yet devised for securing the blocks to one another. A contemporary report speaks of, "… dovetailing, joggles, clamps, bolts, ribbon-band joints, and the most tenacious cement" being used.

The season of 1871 realised a further 14 courses, thus completing the solid trunk of masonry which rose 32 ft 8½ ins out of Chicken Rock and incorporated 1,073 blocks. Above this mark, eight rooms were enclosed by the walls making progress of construction more rapid. Forty-two courses were added during 1872 – a good omen for the following season.

It was 6th June 1873 when the very last stone was fitted into its position on the parapet of the tower, completing a further memorial to the expertise of the Stevenson brothers. Dubh Artach had been brought into service less than a year previously and now a second column of masonry, a mirror image of the Scottish tower, was almost ready to be lit. The interior fitments and furnishings took most of 1874, together with the lantern which was fitted with the best plate glass, "…thick enough to resist the heedless wing of any seabird attracted by the illuminated pane." Fresnel lighting apparatus was installed and two giant fog bells were suspended from the gallery in 1890 to complete the warning apparatus. The total cost was calculated at £64,559.

Towards the end of the year, Chicken Rock lighthouse stood ready to send its warning beams across the swirling waters below. However, there had recently been much correspondence between David and Thomas Stevenson and Trinity House, as to the exact

Above: One of the original hand-painted NLB plans for Chicken Rock lighthouse dated September 1867. Interestingly it shows two towers, one of which has been 'crossed out' and left uncoloured. This bears an uncanny resemblance to Dubh Artach which was currently under construction further north and provided the basis of Chicken Rock's design! The correct tower has been coloured red. *(Northern Lighthouse Board)*

character the light would exhibit. On 29th April 1873, the Board of Trade wrote to Trinity House stating that:

...the character of the light now in course of erection at Chicken Rock should be revolving of the natural colour and attain its greatest brilliancy every half minute and further to that requisite arrangements for throwing from the same lighthouse a red sector of light to mark the path around Langness Point. These arrangements can be carried out either by fixing an auxiliary condensing apparatus, consisting of small holophotes at the rear of the main apparatus in the lantern of the lighthouse or by some other method, which the optical skill of Messrs Stevenson may devise.

The Stevensons were distinctly unhappy about incorporating a red arc and fitting ten holophotes (a lamp with lenses or reflectors to collect the rays of light and throw them in a given direction), in the lightroom, arguing that such additions would require the light to be out of centre in the lantern, ventilation would be poor and the red arc would be of little value in hazy conditions. Trinity House were not to be overruled so easily and stated that such difficulties could be overcome simply by using a larger lantern in the centre of the lightroom and fuelling this with colza oil instead of paraffin. The Stevensons used this design under protest, but preliminary tests confirmed the shortcomings of such apparatus. Exactly what the engineers predicted occurred, forcing an embarrassed Trinity House to concede defeat. The designs originally planned were fitted and these gave the light a range of 18 miles with one flash every 30 seconds. The lighthouse came into service on 1st January 1875, its lighting being the signal for the two towers on the Calf of Man, now obsolete, to extinguish their flames and bring to an end almost 57 years of unbroken service.

The Calf of Man, however, still remained the shore station for Chicken Rock. It was included in a tour made by several of the Commissioners of Northern Lighthouses in 1875 when they visited their newest station. They found that the only other inhabitant of Calf Island was a farmer, and that all supplies had to come from the mainland where the nearest church, school and doctor were also located. The Commissioners were distressed to find that nothing grew in the gardens of the cottage due to, "...legions of rats," and so for humanitarian and financial reasons recommended that the shore station for Chicken Rock be moved to Port St Mary in 1912 where the boatman and his crew who effected reliefs also resided.

Life inside the tower was not nearly as comfortable as the shore station. The keepers on duty had only three rooms that were considered as 'living' accommodation – two bedrooms (with two bunks in each), a living room and a kitchen. All communication with the outside world was by hand signals with an observer who would wait on a headland above Port St Mary at an agreed time each day. Only in 1938 was short wave radio communication finally possible with the rest of the island.

The light from Chicken Rock lighthouse continued to radiate across the turbulent waters around the Isle of Man well into the 20th century without undue incident. However, the most dramatic event concerning this lighthouse occurred comparatively recently, and not, as is often the case, during its construction. It is a story well worth recounting here for it involves the courage and bravery of officers from the Royal National Lifeboat Institution, without whom lives would have undoubtedly been lost on the Chicken Rock.

It was almost Christmas 1960 and the weather around the Isle of Man was typical for the time of year – a cold, crisp winter's day with reasonably clear visibility and a rough sea. At about 11 o'clock on 23rd December the coastguard on Spanish Head, a promontory overlooking the Calf of Man and Chicken Rock, viewed with alarm a pall of smoke rising from the lighthouse. Immediately, the lifeboat station at Port St Mary was contacted and within half an hour Coxswain Gawne and his crew were ploughing their way through the choppy seas aboard R.A. *Colby Cubbin No 2* to investigate such a disturbing event. When the lifeboat reached the reef it was plain that there was indeed cause for alarm. Thick black smoke was issuing from the windows of the lighthouse and billowing upwards while the glass of the lantern was blackened. More horrifying was the sight of the three keepers standing on the reef at the base of the tower underneath a rope dangling from the lantern gallery. They were dressed as if on duty within the warm, dry confines of the lighthouse and not in the waterproofs which were so obviously needed in their present predicament.

Coxswain Gawne was quick to sum up the desperate situation. The tide was rising rapidly and soon the rock on which the three keepers stood would be completely submerged.

Right: The lighthouse on Chicken Rock was built to replace two others on the Calf of Man which were frequently obscured by fog. The lower of the two lights – now in disrepair since its abandonment in 1875 – is seen here as a fast running tide sweeps past the light that replaced it. *(John Hellowell)*

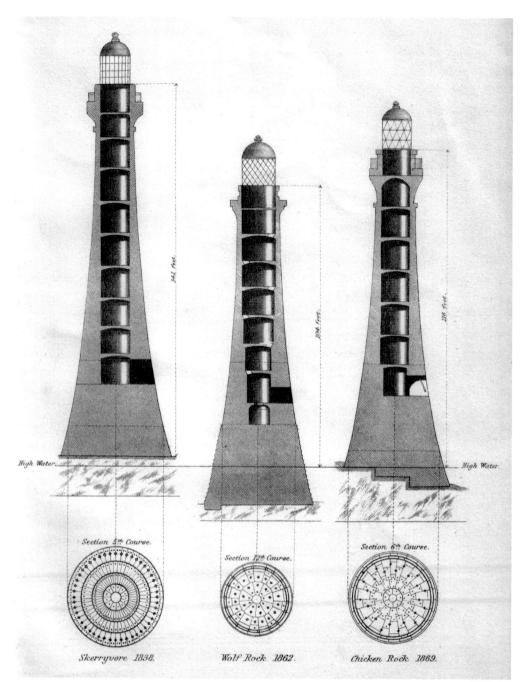

Right: Comparative sections of Skerryvore, Wolf Rock and Chicken Rock lighthouses, showing their positions relative to the high water mark, and also the different methods used by their engineers to joint the blocks of the solid bases.
(from the National Encyclopaedia of 1892)

Not only this, the men, one of whom was clearly in greater distress than the other two, were being subjected to a biting wind which quickly chilled the body to its marrow while their clothing was saturated by the continual assault of waves and spray breaking on the reef. As the tide rose the waves would soon be sweeping across the rock to which the three men were desperately clinging with the risk that all would be swept into the foaming waters and carried away by the swift currents. Adding even further to their peril, above them the interior of the lighthouse was a raging inferno, one room of which contained the large fuel tanks which threatened to explode without warning as a result of exposure to intense heat.

The coxswain had to act with all swiftness if the three men were not to perish under his gaze. He could not approach the reef close enough to enable the men to jump aboard as the rising tide and jagged rocks made such an act impossible. One hundred yards was the closest he could approach. Through a loudhailer he told the men to tie a section of rope around a piece of wood and fling this into the current between the rock and lifeboat, while holding the other end. The wood was immediately swept away from the reef trailing the rope behind it. He then fired a line from the lifeboat which landed across the rope trailing from the reef. By pulling in their rope with the wood attached, the keepers were able to grasp the line from the lifeboat to which a pulley block was attached. This was fastened to the rungs of the dog steps that led from the reef to their lighthouse door, and a continuous rope passed through it. The coxswain was going to attempt a rescue by breeches buoy.

Left: On 23rd December 1960 the Chicken Rock lighthouse suddenly caught fire and forced the three keepers to slide down a rope from the lantern gallery to escape the flames. A dramatic rescue of the stranded keepers then took place by the Port St Mary and Port Erin lifeboats, one of which is seen in this view. Notice how the rising tide means there is now no room at the base of the tower for keepers to shelter. *(Daily Mail)*

The breeches buoy was sent across and the keeper who was suffering the most was helped into the seat by his companions. On the short journey to the lifeboat a huge wave caught the suspended man, toppling him from his seat into the icy waters beneath. Fortunately, the boat crew were able to pluck him from the sea and haul him aboard. He was taken below to be nearer the warmth of the engines and wrapped in blankets. It was obvious to all concerned that medical attention was required rapidly for this man. He was suffering from shock, hypothermia and terrible burns to his hands sustained while sliding down over 100 ft of rope from the lantern gallery.

Coxswain Gawne now made a difficult decision. In view of the keeper's serious condition, and that further rescue by breeches buoy now seemed impossible, he decided to land the injured keeper. He told the two remaining keepers of his intentions over the loud hailer and that he was requesting the Port Erin lifeboat to stand-off the lighthouse in his place. This done he manoeuvred out of the tricky currents around the rock and made all speed for Port St Mary. The two stranded keepers were left to ponder their fate. The rising tide was creeping perilously close to their feet, curtains of spray rained down on them soaking every last stitch of clothing, and the lighthouse was threatening to explode above their heads. With the lifeboat rapidly disappearing into the distance, these men must have indeed wondered if they would spend Christmas 1960 with their families.

However, unbeknown to the two keepers, a great deal was being done to rescue them. The Port Erin lifeboat was on its way and an Air-Sea Rescue helicopter had arrived from RAF Valley on Anglesey and was circling the smoking beacon. It was obvious to the two keepers that rescue by this method was impossible as the helicopter could not approach sufficiently close for the men to be winched aboard for fear of the rotor blades touching the tower. Only a rescue from the lantern gallery was feasible by this method but the men could clearly not be expected to climb back up the rope. This being the situation, the helicopter returned to base and left the keepers alone once more.

It was not long before the Port Erin lifeboat *Matthew Simpson* arrived to stand-off the reef. By three o'clock that afternoon the Port St Mary boat had landed its casualty and was returning to Chicken Rock. The two boats bravely stood by in the hope of attempting a rescue. As the daylight began to fade, the seas began to moderate and the tide turned. If a rescue was to be attempted now was the time as the fast approaching darkness would prevent any further action until the morning. Darkness was no time to be manoeuvring a small craft around the jagged teeth of Chicken Rock.

Right: The damage inside the tower was repaired by using gunite – a new type of sprayed concrete. All the machinery was brought in by helicopter which set down on a specially constructed helicopter platform seen on the right of this reef.
(*Manx Press Pictures*)

The calmer sea and receding tide made an impossible situation slightly easier. Gawne edged his craft, inch by inch, towards the desperate men. The rocks that were previously covered were slowly appearing so he could gingerly pick his way through them. Eventually, he was close enough for the two exhausted keepers to leap aboard and so end an eight-hour ordeal. They were burnt, shocked and weak from their horrifying experience – but they were alive. While they were revived below the coxswain edged his boat from under the towering inferno. It was but a few short minutes before the two men were reunited with their families, an occasion made possible only by the exceptional courage and bravery of the lifeboat crew, character traits that are so often found in this dedicated breed of men.

Later that day the Admiralty issued an exceptional notice to shipping. It read, "the Chicken Rock light has been extinguished and the fog warning is not, repeat not, operating." When the three keepers were sufficiently recovered to tell of the day's events it was discovered that at about half-past ten on the morning in question the lighthouse was rocked by an explosion in one of the lower rooms. This started a fire which quickly spread, fuelled by the interior wooden fitments, and drove the men upwards on to the lantern gallery with no time to protect themselves against the elements. The fire continued to rage ever upwards through the tower, but for the men on the gallery there was no further upward retreat from the advancing flames.

The short wave radio was below them in the burning tower so couldn't be used to summon help. Five distress rockets were fired to attract the attention of a passing ship, but they went unseen. Two shots were fired from the fog gun which were successful. They alerted the relief keeper at the shore station an he summoned the Coastguard. Below them though, the flames were steadily creeping upwards. If they were to survive there was only one course of action. They made use of a rope which was tied to the balcony railings and trailed downwards, thus effecting a slow, painful but only temporary escape. Having evaded the flames they were now at the mercy of wind and wave. Fortunately, subsequent events proved their actions in abandoning the lighthouse to be for the best.

Left: A dramatic view of Chicken Rock taken from the landing during a routine maintenance visit in 1996
(*John Hellowell*)

Two days later, as the last few wisps of smoke curled away from the gutted tower, the Commissioners of Northern Lighthouses fixed a flashing buoy some distance off the reef from their relief vessel *Hesperus* which had steamed from Oban. By 29th December a further improvement was made by installing an unwatched light in the tower of the old lighthouse on the Calf of Man while they considered what to do about their unserviceable lighthouse. The winter weather prevented a detailed examination from being made until several weeks later. Only after a complete survey could any permanent repair scheme be considered. Here it will be remembered that a similar situation occurred at Skerryvore lighthouse when it was damaged by fire in 1955. The success of this repair using gunite, a type of sprayed concrete, led the Northern Lighthouse Board to adopt a similar method at Chicken Rock, although it was realised that the repair to Chicken Rock would present greater difficulties for several reasons.

Left: The new lantern inside Chicken Rock is remarkably small compared to the original apparatus. Outside, on the gallery, NLB personnel carry out routine safety checks on the lantern roof. (*John Hellowell*)

Firstly, the extent of the damage was far greater than at Skerryvore and was going to require a great deal more capital, not to mention expertise, to put it right. Secondly, unlike Skerryvore, there is no room outside the lighthouse on which the extensive and bulky equipment required for the process could be sited. The reef is completely covered at high tides so every piece of machinery would have to be of such a size that it could fit through the lighthouse door to be stored inside. This led to pieces of equipment being specially designed for the purpose. Lastly, to overcome the problem of landing men, materials and food, etc, it was decided to construct an aluminium helicopter landing platform on the reef, a decision prompted by the uncertainty and difficulties of landing on Chicken Rock from a boat. The vagaries of the weather, difficult approach and swift currents made landing just men a hazardous undertaking. With heavy equipment and materials such difficulties would be magnified. A hovercraft was even considered for this role but in the end it was wisely decided that a helicopter would present the most reliable and efficient method of transport from the base. It was a judgement they were not to regret.

The amount of detailed pre-planning was enormous. The same contractors who repaired Skerryvore were engaged – a Liverpool firm, Whitley Moran & Co Ltd, experts in the use of gunite. A Bell 47 helicopter was chartered and fitted with floats as required by the Ministry of Civil Aviation, and a local fishing boat was also engaged as a back-up vessel. All these and similar details took over a year from the date of the fire.

By May 1962 attempts were made to land keepers on the lighthouse to clear the interior debris, establish a radio telephone link with the base at Port St Mary, and erect the helicopter landing platform. This took two weeks, the contractors then took over with one keeper remaining as cook and radio operator. Internal repairs were tackled

Below: Early morning sunlight glints on the the side of the NLB tender *Pharos* as she drifts past Chicken Rock with the NLB helicopter on her stern. (*Eddie Dishon*)

first. It was found that some of the cracks in the stonework were up to 9 ins deep. The damaged granite was cut out by pneumatic chisels and replaced by gunite, sprayed from a compressed air gun. Stainless steel reinforcement was used in some areas but not where the danger of corrosion was likely. Once the internal rooms were completed, two steeplejacks inspected the exterior walls

Left: Almost flat calm as *Seacat Isle of Man* from Douglas to Belfast powers past Chicken Rock lighthouse *(Northern Lighthouse Board)*

from a cradle suspended from the lantern gallery. Any damaged joints were repointed under pressure although this part of the work was halted several times by high winds making work on the cradle impossible.

The use of a helicopter proved invaluable. Three tons of plant and equipment, 25 tons of sand and cement, as well as regular supplies of food and drinking water all arrived promptly by helicopter. Landings were made in the two hours either side of low water unless high winds or spray washing over the landing platform made it dangerous to do so. Even so, there were very few days on which flying was impossible and the helicopter was able to keep the men in the lighthouse continually supplied with all their requirements. Supervising engineers from the Lighthouse Board were able to visit regularly and on several occasions urgently-required spare parts from Liverpool were able to reach the lighthouse within 24 hours of being requested. All this was in great contrast to the work on Skerryvore where prolonged delays of getting materials and food were experienced because of all transport being by boat. The helicopter was in fact able to supply Chicken Rock at a faster rate than the men could hoist the goods into the tower. Only 12 minutes were required for a round trip from the base to the lighthouse, including unloading time.

Below: A poignant view of Chicken Rock lighthouse through the smashed lantern panes of the lighthouse it replaced on the Calf of Man. *(John Hellowell)*

It was estimated that six weeks would be required to complete the work. It took exactly that time. On 10th September 1962 a notice was issued to mariners stating that on or about 20th September a permanent unwatched light – powered by acetylene – would be established at Chicken Rock 125 ft above MHWS and that also a temporary unwatched electric fog horn would be operative from the same date. The Chicken Rock lighthouse would never again have lightkeepers on a regular basis. Thirty seven years later, on 24th March 1999 the lighthouse completed another conversion, this time to solar electric operation.

The lighthouse is still a prominent feature off the southern tip of the Isle of Man although how many of the tourists who gaze at its shapely form are aware of the near tragic events of December 1960? It is only due to the bravery from the crew of the Port St Mary lifeboat that the three keepers lived to tell of such events.

Opposite: An atmopsheric view of Chicken Rock at dusk. *(John Hellowell)*

14 FLANNAN ISLES
The Marie Celeste lighthouse

FAR, FAR OUT in the grey wastes of the Atlantic Ocean, 20 miles further west than the Outer Hebrides, almost 80 miles from the Scottish mainland, are a cluster of islands known collectively as the Seven Hunters or Flannan Isles. Their total area is less than 100 acres, formed by seven larger islands interspersed with considerably smaller islets and skerries that stir the Atlantic tides into confusion as they race between the group. They are a favourite breeding ground for large colonies of puffin, petrel, fulmar, shag and kittiwake, but are uninhabited by man. However, this has not always been so. Ancient records show these islands to be the home of St Flann, an obscure saint who is reputedly the cause of the many superstitions surrounding his domain. The religious significance of these islands became so marked that prayers were supposedly said by all who intended to land there as a protection against ill-fortune. In more recent times the Flannan Isles were a favourite grazing spot for the shepherds of Lewis who would transport their sheep by boat to graze on the lush cliff-top grass.

Left: The Flannan Isles lighthouse as seen from the approaching NLB helicopter.
(Northern Lighthouse Board)

The larger islands of the group have retained their gaelic names. Here are found Eilean a'Ghobha, Eilean Tighe and Eilean Mor, but for this chapter we concern ourselves only with the last named and largest of these islands, Eilean Mor. During the closing years of the last century a lighthouse was built here by David and Charles Stevenson. Eilean Mor possesses some 29 acres above the water and at its highest point is 285 ft above sea level. There was obviously no difficulty in siting the lighthouse on the summit of the island although the landing of materials for construction proved not such an easy task. Eilean Mor rises sheer from the Atlantic. Landing is difficult even in the calmest of conditions, so two giant flights of steps to scale the cliffs were constructed from concrete blocks on either side of the island. This enabled boats to moor and unload at varying states of wind and tide with a great deal more ease than was previously the case. With these steps, together with landing cranes and a railway track, it was possible to transport men and materials from boats to the lighthouse site.

The light took four years to construct between 1895 and 1899. The tower rises 75 ft, giving the lantern an elevation of 330 ft above the waves and a range of 25 miles. It was first

Opposite: The NLB helicopter lifts off from the Flannan's helipad. Rows of solar panels that now provide all the power for the lighthouse have been erected on a new gantry.
(Peter Christmas)

lit on 7th December 1899 when the engineers and workmen deserted the island leaving only the three keepers in the light.

From such a brief account it can be correctly assumed that there was nothing unusual or dramatic about the construction of this lighthouse. It was just another remote Scottish island whose position was now clearly marked by a lighthouse, the building of which bears no comparison with some of the earlier efforts of the Stevenson family. Had it not been for the events which occurred here in 1900 the name of the Flannan Isles would doubtless have slipped into oblivion, unknown to all except the local mariners, officers of the Lighthouse Board, or those with a keen interest in lighthouses. However, late in 1900 the Flannan Isles were the scene for a chain of disturbing events which have resulted in one of the most remarkable sea mysteries ever told. The account which follows contains only facts; facts which have produced a tale of tragedy and intrigue which has still not been satisfactorily resolved to this very day.

The Flannan Isles' mystery begins only a few days after the first anniversary of the lighthouse's construction. The *Archtor* (also referred to as the *Archer* in some reports), a steamer from Philadelphia to Leith, passed close to the Flannans on 15th December 1900 yet her captain was unable to pick up the light he knew should be there. The weather was not good, but visibility was such that the silhouette of the islands could be picked out without undue difficulty. There was no flash from the summit of Eilean Mor, this fact being reported after docking at Oban. Curiously, nothing appears to have been done at this stage.

By fortunate coincidence, relief of the keepers was due on 20th December, although a series of gales prevented the relief vessel *Hesperus* from leaving her berth on Little Loch Roag on the west coast of Lewis until 26th December. By noon the *Hesperus* was in sight of the islands and soon anchored off the eastern landing. On board were Captain Jim Harvie, second-mate McCormack, Joseph Moore who was starting his turn of duty at the light, and Allan Macdonald, a buoymaster in charge of the routine inspection and servicing of buoys en route, plus a normal complement of seamen.

As soon as the boat had dropped anchor it was obvious to the crew that all was not as it should be. The flagstaff, where a flag was generally flown on relief days, was bare. The landing stage should have been piled with provision boxes to be taken for refilling. None were to be seen. Most obvious

Left: An interesting aerial view of Eilean Mor showing the lighthouse and the tramways that lead to the east (on the right) and west landings. *(Northern Lighthouse Board)*

was the lack of any inhabitants waiting to greet the relief vessel. After a month on the lonely Flannans the keepers were eager to learn of news from the mainland and normally waited expectantly by the landing stage. This time there was absolutely no sign of any life, even through a telescope.

Thinking the keepers might have mistaken the date and were unaware of their presence, the captain let go a strident blast on the steam whistle. It came bouncing back off the cliffs of Eilean Mor. Another blast was sounded which was greeted as before – with nothing. A distress flare was fired into the sky above the lighthouse, bursting into vivid colour above the island. Surely they must see that. Nothing. There was obviously something seriously amiss. A boat was lowered containing Joseph Moore, second mate McCormack and a few seamen, which soon covered the distance between the boat and the island. Joseph Moore was the only one to land and he made his way up the steep flights of steps to the lighthouse.

He found the entrance gate closed, as was the entrance door and the door inside that. The kitchen door was open and through it Moore could see that the fire had not been lit for several days and that the clock had stopped. In the bedrooms the beds were unmade and empty. In a letter about the events Moore states that, "I did not take time to search further, for I only too well knew that something serious had occurred." He ran down the precipitous steps and told McCormack of his findings. Both men, along with a seaman, then returned to the light to confirm the story. In the lightroom they found everything in its proper place; the lamps had been cleaned and refilled and the blinds were drawn around the perimeter, although the daytime cloth had not been draped over the lens – a serious breach of regulations. They could see no reason why the light should not be operating.

Downstairs an uneaten meal of cold meat, pickles and boiled potatoes was allegedly on the table, at the side of which was an upturned chair. There is some doubt about the authenticity of this 'meal' as a later statement by Joseph Moore, the first independent observer of the scene, contains the observation that, "The kitchen utensils were all very clean, which is a sign that it must be after dinner some time they left." W.W.Gibson's poem about the events, a verse from which I reproduce at the end of this chapter, contains references to a table being spread with an untouched meal – the probable source of this popular misconception.

Outside the lighthouse no sign could be found of any of the keepers. They had, for no apparent reason, vanished into thin air. It was a frightening situation for the men, who quickly returned to their boat to relay the distressing news to Captain Harvie. He advised Moore and buoymaster Macdonald, together with two volunteer seamen named Lamont and Campbell, to return to the island and tend the light while he returned in the *Hesperus* to Breasclete on Lewis to send a telegram to the secretary of the Northern Lighthouse Board. It read as follows:

Left: The west landing on Eilean Mor showing the huge flights of steps that had to be constructed, together with a small tramway, to transport supplies to the lighthouse on the summit of the island. The photograph was taken in 1900 shortly before the three keepers disappeared, probably swept away by a giant wall of water from this very landing.
(F. S. Thompson)

A dreadful accident has happened at Flannans. The three keepers, Ducat, Marshall and the Occasional have disappeared from the Island. On our arrival there this afternoon no signs of life was to be seen on the Island. Fired a rocket but, as no response was made, managed to land Moore, who went up to the Station but found no Keepers there. The clocks were stopped and other signs indicated that the accident must have happened about a week ago. Poor fellows they must have been blown over the cliffs or drowned trying to secure a crane or something like that. Night coming in, we could not wait to make further investigation but will go off again tomorrow morning to try and learn something as to their fate. I have left Moore, Macdonald, Buoymaster and two seamen on the Island to keep the light burning until you make other arrangements. Will

not return to Oban until I hear from you. I have repeated this wire to Muirhead, in case you are not at home. I will remain at the telegraph office tonight until it closes, if you wish to wire me."
Master, Hesperus

Ninety minutes later the telegram boy delivered the message to the Board's offices in George Street, Edinburgh. Unable to get a reply he took it to the Board's secretary, William Murdock, who was somewhat put out at receiving a telegram at such a late hour, but read it nevertheless. He was stunned by its contents.

The following day the four new keepers of the light combed the island for clues. They found precious few. At the eastern landing place everything was as it had been left after the previous relief of 7th December. The ropes were still neatly coiled and in their place. At the western landing, however, there was dramatic evidence as to the severity of the recent weather. A wooden box used to store ropes and crane handles had been ripped from its position in a crevice of the rock 110 ft above sea level. Its contents were strewn about haphazardly, but the ropes were still coiled – indicating they had not been used. Also, the iron railings surrounding the landing crane were bent in an alarming fashion, and a large block of stone weighing in excess of a ton had been displaced at an even higher level. A life buoy, fastened to railings alongside a concrete path, was missing, torn from its fastening ropes by the sea leaving only shreds of canvas still attached to its securing ropes.

Further demonstrations of an ocean's fury were seen along the cliff-top where the turf had been swept away for a distance of 30 ft. This was 200 ft above the high water. It is difficult for mortals not intimately connected with the sea to imagine what it must have been like on the Flannan Isles during the storm which caused this damage.

Part of Muirhead's report for the NLB reads as follows;

Left: The size of the optical apparatus inside a modern lighthouse can be comparatively small when compared to the lantern room in which they are housed. This is the case at the Flannan Isles lighthouse after automation. The silhouettes of visiting engineers are also seen in the lantern room.
(The Glasgow Herald)

On the Thursday and Friday the men made a thorough search over and round the island and I went over the ground with them on Saturday. Everything at the east landing place was in order and the ropes which had been coiled and stored there on completion of the relief on the 7th December were all in their places, and the lighthouse buildings and everything at the station was in order. Owing to the amount of sea, I could not get down to the landing place, but I got down to the crane platform about 70 feet above the sea level. The crane originally erected on this platform was washed away during last winter, and the crane put up this summer was found to be unharmed, the jib lowered and secured to the rock, and the canvas covering the wire rope on the barrel securely lashed round it, and there was no evidence that the men had been doing anything at the crane. The mooring ropes, landing ropes, derrick landing ropes and crane handles, and also a wooden box in which they were kept and was secured in a crevice in the rocks 70 feet up the tramway from its terminus, and about 40 feet higher than the crane platform, or 110 feet in all above sea level, had been washed away, and the ropes were strewn in the crevices of the rocks near the crane platform and entangled among the crane legs, but they were all coiled up, no single coil being found unfastened. The iron railings round the crane platform and from the terminus of the tramway to the concrete steps up from the west landing, were displaced and twisted. A large block of stone weighing upwards of 20cwt, had been dislodged from its position higher up and carried down to and left on the concrete path leading from the terminus of the tramway to the top of the steps. A lifebuoy fastened to the railings along the path, to be used in case of emergency had disappeared, and I thought at first it had been removed for the purpose of being used but, on examining the ropes by which it was fastened, I found that they had not been touched, and as pieces of canvas were adhering to the ropes, it was evident that the force of the sea pouring through the railings had, even at this great height (110 feet above sea level), torn the life buoy off the ropes.

Above: A team of NLB engineers are at work in the lighthouse while the NLB helicopter that brought them is shut down on the helipad to the right.
(P. D. Green)

It was obvious, however, from evidence in the lighthouse, that all three keepers had survived this bombardment. The log book was kept by the principal keeper, Ducat. In it entries were made up until 13th December, and on a slate alongside it, ready to be written up in the log, were the details for 14th December and up until 9 am on 15th December. These logs are to be found at all lighthouses and record such information as weather conditions, state of the sea, wind speed and direction, plus other relevant notes such as barometer readings and the times of lighting and extinguishing the lamp. The log had detailed the poor weather conditions which had prevailed during most of December and showed that they had withstood the evil weather which had damaged the fittings on the west landing.

The kitchen utensils were all cleaned and put away, indicating that the person who was cook for the day had completed his work. It is therefore assumed that the men disappeared on the afternoon of Saturday, 15th December, an assumption borne out by the fact that the steamer *Archtor* failed to pick up the Flannan's light that evening. No clues had yet been found to explain why all three men had disappeared.

On 29th December, Robert Muirhead, a superintendent for the Northern Lighthouse Board, arrived on the scene with Donald Jack and John Milne, two experienced keepers who were to take over the manning of the lighthouse with Joseph Moore. On arrival he was shown the clothes locker and its contents of one set of oilskins, a reefer jacket and a pair of sea-boots. Moore, who was familiar with the normal outdoor wear of the keepers, explained that these remnants showed that James Ducat's cape and boots had gone (Ducat preferred a cape to oilskins), so had Thomas Marshall's boots and oilskins. The remaining oilskins, jacket and boots were therefore Donald McArthur's. Moore further confirmed this from his knowledge that McArthur never used any other coat apart from his old reefer jacket, and he therefore assumed that McArthur left the light in a hurry in his shirt-sleeves. He also knew that the only time Marshall wore his boots and oilskins was when he was going down to the landing stage. The weather entry in the log for 15th December showed heavy rain – requiring the use of waterproof clothes if any of the men were to leave the light. Why then were McArthur's jacket and boots still in the locker?

These were the scraps of information gleaned from an exhaustive search of lighthouse, island and landing stages. They indicate clearly that all three men left the lighthouse on the afternoon of 15th December, for an unknown reason, and failed to return. The light was unattended from this date until the 26th when the relief boat arrived. There were no positive clues to explain exactly why all three men should leave their posts, presumably in enough of a hurry to cause an upturned chair, especially as one man should have remained in the building at all times. No bodies were ever found.

Once the tragic circumstances became widely know, rumours to explain the disappearance became rife. Amongst the more colourful was that one of the keepers killed his two companions and disposed of their bodies over the cliffs. He was then overcome with remorse and took his own life in a similar fashion. It is difficult to believe that serious-minded folk put forward the idea that a sea-serpent had snatched the men from their lonely home or that they were plucked from the island by a giant marauding sea-bird. Kidnapping by foreign agents was mentioned, as was the possibility of a giant wave sweeping over the Eilean Mor taking the unfortunate men with it. It was, of course, not difficult to pick fault with all such ludicrous suggestions.

Below: A view of several of the Flannan Isles from an approaching boat. This is a very similar view to the one that the Captain and crew of the *Hesperus* would have seen in December 1900 as they investigated the disappearance of the three keepers.
(John Cunningham)

To the deeply religious and superstitious families of the Western Isles one theory which merited deeper consideration concerned the 'Phantom of the Seven Hunters' – a nebulous being, reputedly having some connection with St Flann, who kept watch over these islands and was greatly displeased by the intrusion into his solitude by a lighthouse and keepers. The disappearance of the men was his retribution. Swiftly dismissed as pure fantasy by those living away from the Hebrides, local people were quick to point out that although only a year old, the lighthouse had lost three keepers, another had fallen to his death from the lantern gallery, while four men were drowned when their boat overturned approaching a landing stage. Coincidence? Probably, but nevertheless an unsettling chain of events in such a short period.

The official explanation produced by Robert Muirhead is a little more down to earth. Part of it reads:

After a careful examination of the place, the railings, ropes, etc, and weighing all the evidence which I could secure, I am of the opinion that the most likely explanation of the disappearance of the men is that they had all gone down on the afternoon of Saturday, 15th December to the proximity of the West landing, to secure the box with the mooring ropes, etc, and that an unexpectedly large roller had come up on the Island, and that a large body of water going up higher than where they were and coming down upon them had swept them away with resistless force.

I have considered and discussed the possibility of the men being blown away by the wind, but, as the wind was westerly, I am of the opinion, notwithstanding its great force, that the more probable explanation is that they have been washed away as, had the wind caught them, it would, from its direction, have blown them up the Island and I feel certain that they would have managed to throw themselves down before they had reached the summit or brow of the Island.

This is certainly a reasoned and considered judgement but it does not surely explain the overturned chair or why the third man had left the lighthouse without waterproof clothing at a time when it was raining heavily. They would not have left a meal on the table to go cold, if there ever was such a meal, and rushed out of the room, upsetting a chair, to undertake a task that could equally well have been done later when the weather may have moderated.

There is, however, one explanation which appears to go some way to unravel the Flannan Isles mystery. It was put forward by men who regularly visited the island with their sheep, men who knew the peculiar wave forms and natural contours of Eilean Mor better than most. More importantly, it could explain the clues left on the island after the keepers had disappeared – something the official explanation had failed to do. The coastline of Eilean Mor is indented at several points by deep, narrow gullies called geos, into which the waves can travel unimpeded. The west landing stage was actually set in one of these geos known as Skiopageo, a narrowing creek some several hundred yards in length and terminating in a cave which became completely filled at high tide. During times of storm the waves would rush into the inlet and compress the air in the cave, causing it to explode outwards sending curtains of spray and considerable volumes of water into the air. It was suggested that during the heavy seas on 15th December, Ducat and Marshall went to check on the security of equipment at the west landing leaving McArthur to continue with the preparation of the meal. While doing this his attention was caught by a series of exceptional waves heading for the island from the open Atlantic. He well knew the force with which the cave at the end of Skiopageo could discharge its watery contents and so ran to warn his fellows of the approaching danger, leaving his waterproofs and knocking over a chair in his haste. By the time he reached the others it was too late and all three were swept into the sea by the sheer weight of descending water.

This, then, was the final and most plausible explanation for the mysterious events. Once they became known to the Press, stories without number were published purporting to be the key to the mystery. Only the one detailed above merited consideration, most were the product of a colourful imagination. To this day no further clues have been found to explain why three loyal and competent keepers should vanish from their post leaving precious little to explain the true course of events. It seems that the Flannan Isles will keep their secret into eternity and remain the focal point for what must surely be a unique lighthouse story.

The lighthouse today is an unmanned station. During 1971 work went ahead to convert the lighthouse to an automatic system of operation. A helicopter pad was built about 75 yards from the buildings which enabled the conversion work to be undertaken swiftly. The old lens and mechanism were replaced by their modern equivalents, the illumination being provided by acetylene gas, over a year's supply being stored on the island. The keepers were finally withdrawn on 28th September 1971. Further modernisation was achieved in 1999 when the lighthouse has converted from acetylene gas to solar electricity and wind generators. The new light operated from 24th March and its operation is monitored from NLB headquarters in Edinburgh.

It is fitting that the events of that fateful December have been immortalised in a poem by W.W. Gibson entitled *The Flannan Isles*, that has been inspired by these mysterious islands. Its last verse chillingly reminds us of the situation the NLB personnel must have found themselves in after landing on the island in December 1900:

> *We seemed to stand for an endless while,*
> *Though still no word was said*
> *Three men alive on Flannan Isle,*
> *Who thought on three men dead.*

Below: Eilean Mor in all its splendid isolation. *(Rod Huckbody/Stornoway Gazette)*

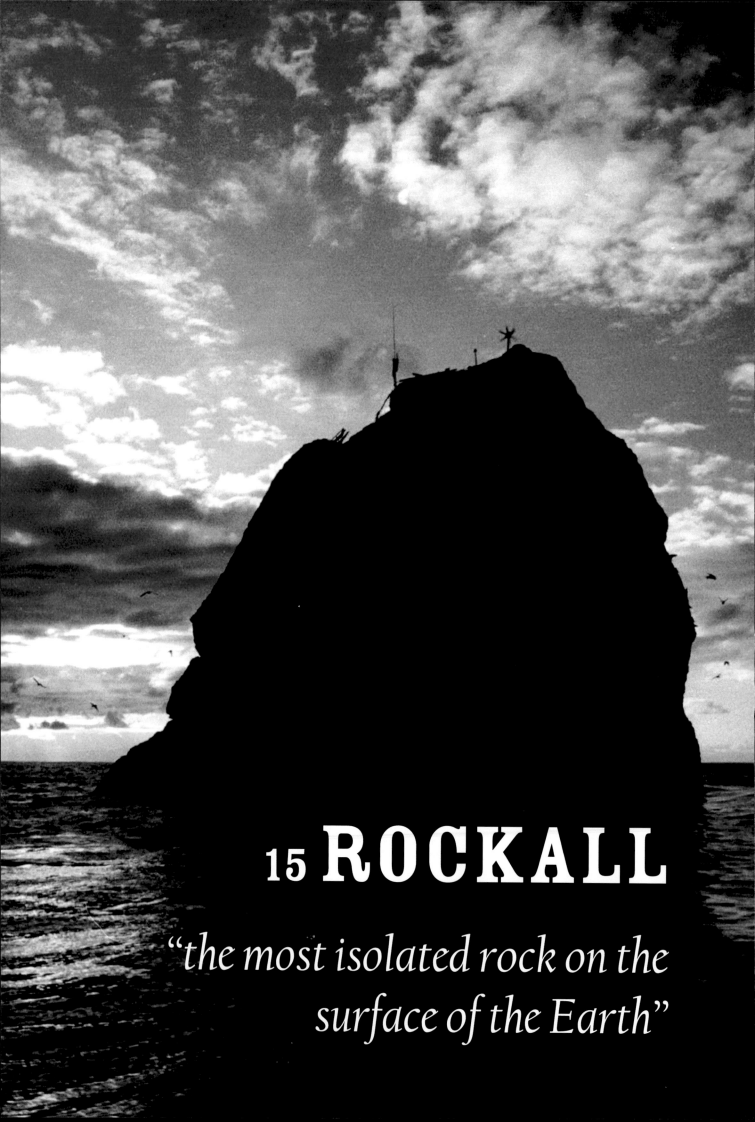

15 ROCKALL

"the most isolated rock on the surface of the Earth"

FOR THIS final chapter I've chosen a navigational aid that could hardly be described as a 'lighthouse' in the accepted sense of the word. It doesn't have a tower with internal accommodation and a lantern on top like Eddystone, Skerryvore and the rest. Nor does it have keepers' dwellings adjacent to a tower like the Skerries or Muckle Flugga. It's never needed them, because the beacon on this particular rock has never had keepers. In fact, it's only had its light for a little over 30 years. But, as rock 'lighthouses' go, this one is in a league of its own. What sets it apart from any of the others you have read about so far, is its location. If we regard the isolation of a particular rock as being measured by how far it is from the next nearest rock or landfall, then this is the daddy of them all! This is Rockall, once described as "the most isolated rock on the surface of the earth" – and the beacon on its summit is the most isolated British 'lighthouse' of all.

Rockall is a lonely lump of granite that breaks the surface of the North Atlantic so far to the west of the British Isles that it rarely features on any map. There are only two maps that can be relied upon to show you exactly where Rockall is. One is the map of the shipping forecast sea areas that are so valuable for offshore shipping around Britain. Rockall has given its name to one of these areas – an almost rectangular block of Atlantic Ocean. Most are named after the adjacent landfall, but Rockall takes its name from a feature so small it's impossible to mark on the actual shipping forecast chart.

The other place you'll find Rockall is on the official Admiralty chart for that particular section of wild Atlantic. Again the rock is so small its location is only betrayed by a small black dot and its name. Basil Hall, the man who is generally accepted to be the first person to land on and climb Rockall in 1811, wrote after the event that, "This mere speck on the surface of the waters – for it seems to float on the sea – is only 70 feet high and not more than a hundred yards in circumference. The smallest point of a pencil could scarcely give it a place on a map which should not exaggerate it in proportion to the rest of the islands in that stormy ocean."

Opposite: Rockall in tranquil mood as the sun sets behind it. On the summit is a wind generator brought by the Greenpeace activists who occupied it in 1997. *(Greenpeace - Sims)*

Above: Rockall from the east in September 1955 as a heavy swell smashes against the rock. *(Natural History Museum)*

Let's put this location into some kind of perspective. The 'official' Admiralty position is 57° 35′ 48″N, 13° 41′ 19″W. That places it 250 miles off the west coast of North Uist in the Outer Hebrides – and 235 or so miles to the south west of the Flannan Isles. The nearest point on the Scottish mainland is the lighthouse at Ardnamurchan Point – 287 miles away. It's 265 miles from the nearest point of the Irish Republic in County Donegal, just over 400 miles from Torshavn, the capital of the Faroe Islands which rise from the same submarine ridge as Rockall, and finally, it's 460 miles from the nearest landfall on Iceland. The significance of these distances will become clear later in the chapter. Its description of being "the most isolated rock on the surface of the Earth" is probably true – and it's British! We don't hold many world records when it comes to physical features on our planet – but this is one of them.

So exactly how small is Rockall? It has actually been surveyed on several occasions, revealing that it's between 80–100 feet wide across its base and then rises, roughly in the shape of a cone or triangle, to a height of about 72 feet. Its eastern face is sheer, while its western and southern faces are not far off vertical. There's no obvious landing site and any attempt to access Rockall is an extremely dangerous and risky manoeuvre that can only be attempted in the calmest of seas. Again, some perspective on its size; Bishop Rock's tower (the tallest of Trinity House's rock lights) is 160 feet high, while the NLB's tallest, Skerryvore, is 138 feet. Stood next to one of these giants Rockall wouldn't reach half way up Bishop Rock and just creep above the half way mark of Skerryvore. For a rock in the middle of the wild Atlantic that is small, so it's no surprise to find that during severe winter storms, waves and spray sweep completely over the top of Rockall.

It is actually the remnant of an extinct volcano, a plug of solidified lava that erupted 40 million years ago and has resisted the subsequent millions of years of erosion. When first analysed the rock was found to be a coarse mix of granite and quartz that was rich in soda, together with traces of all sorts of rare minerals. This was called rockallite by its earliest surveyors, who also discovered that it's highly magnetic, an effect felt some two to three miles off the rock, and therefore a particular problem for ship's compasses.

Rockall has no soil and no plant life – apart from microscopic algae and one species of black lichen. Only six species of animal live there permanently, and you would be hard pressed to find any of them if you were ever fortunate enough to set foot there. Two species of rough periwinkle, two species of common mite, an orange rotifer and a trematode all cling to Rockall underneath a skirt of seaweed that marks the waterline.

Rockall does, however, provide a temporary respite for over twenty species of sea bird. Nesting and breeding of these species is considered unlikely because there is little material available to make their breeding 'mounds' or nests, and it would only take one wave to sweep over Rockall to carry everything on the rock into the Atlantic. But even as a resting 'perch' such a large number of species do produce a considerable quantity of droppings – otherwise known as guano. From a distance the white colouring around the top of the rock is said to look like the approach of a ship in full sail.

Exactly when Rockall was discovered is shrouded in mystery. There are reports that

Left: The sheer eastern face of Rockall shows up well in this view of Rockall in fairly choppy seas.
(Natural History Museum)

a Saint Brendan landed there from Ireland in the sixth century, but this cannot be confirmed and is unlikely to be true. Its huge isolation inevitably gives rise to all sorts of myths and legends, and the Irish were not slow to attribute its being to the work of their mythological giant Finn McCool who is alleged to have thrown a 'pebble' into the sea which subsequently turned out to be Rockall. Norse warriors who first navigated the waters around the north and west of Scotland had on their charts a place called 'Friesland' meaning an area of low shoals, in roughly the same area that Rockall is today.

However, its first reliable appearance on a map was not until 1606 when it appeared on a chart produced by Willem Jansen as 'Rocol', at a location that turned out to be only 87 miles from its true position. In 1703 a book called *A Description of the Western Isles of Scotland* was produced by the interestingly named Martin Martin. A map in this volume shows the St Kilda group of islands clearly, while further west still is a feature marked 'Rokol'. Over the next century it appeared on a succession of different charts from different countries as either 'Rochol', 'Rockol', 'Rookal' or 'Rokele' although not in a constant position. Captain W. Coats bound for Liverpool from Hudson Bay is credited with being the first to accurately chart its position in 1745. He fixed it at 57° 38′N, 13° 55′W – just eight miles from its actual location.

Left: The Danish liner *S.S. Norge* before she met her tragic end close to Rockall in June 1904 with the loss of over 600 souls.

The appearance of Rocal or Rokol gives us some clues as to how it got its name. It's generally accepted to be Gaelic in origin – 'sgeir rocail' was its original name; *sgeir* being gaelic for 'rock' (compare this with Skerryvore's origins) and *rocail* meaning either 'roaring', 'tearing' or 'ripping' depending on which translation you subscribe to. Hence, we end up with a name by which it was reputed to be known to Hebridean sea traders "from time immemorial" as 'the sea rock of roaring'. There's also no doubt that 'ripping' and 'tearing' would also be highly appropriate adjectives for this vicious rock – vividly demonstrated on 22nd August 1686 when a Spanish merchant ship bound for the 'New World' "ran on the rocks" at Rockall with the loss of 250 lives.

On 8th September 1811 one of the officers on board HMS *Endymion* was Lieutenant Basil Hall. They were sailing the waters west of the Outer Hebrides when the look-out in the fore-topmast-head reported a sail on the lee side. None of the crew could make out what they were looking at, so the *Endymion* gave chase. At this time, it must be remembered that Britain was at war with France so all unknown vessels approaching British waters were investigated. Hall recorded the events of that day in his memoirs – six volumes collectively entitled *Fragments of Voyages and Travels Including Anecdotes of a Naval Life.*

The general opinion was that it must be a brig with very white sails aloft, while those below were dark, as if the royals were made of cotton and the courses of tarpaulin – a strange anomaly in seamanship it is true, but still the best theory we could form to explain the appearances. A short time served to dispel these fancies, for we discovered on running close to our mysterious vessel that we had actually been chasing a rock – not a ship of oak and iron, but a solid block of granite – growing, as it were, out of the sea at a greater distance from the mainland than, I believe, any other island, or islet, or rock of the same diminutive size to be found in the world.

Its name is Rockall and it is well known to those Baltic traders which go north-about. We were deceived by it several times during the same cruise, even after we hade been put on our guard, and knew its place well. As we had nothing better on our hands, it was resolved to make an exploring expedition to visit this little islet. When we left the ship the sea appeared so unusually smooth that we anticipated no difficulty in landing, but on reaching the spot we found a swell rising and falling many feet, which made it an extremely troublesome matter to accomplish our purpose.

He probably didn't realise it at the time but Hall was about to attempt exactly the same feat as all the builders of the great rock lights around the British coast had done already, or would do in the subsequent years, namely, to clamber on to the rock or reef where their proposed tower would be built for the very first time. Winstanley, Smeaton, Whiteside, Walker, the Stevensons and the Douglass's all had to make that first leap from an open boat onto their reef, a tricky and dangerous manoeuvre requiring split second timing, nerves of steel and the sure-footedness of a mountain goat. Launching yourself from a pitching boat onto a seaweed and

Right: Sgt. Peel and Cpl. Fraser bolt the brass plaque annexing Rockall for Britain on to Hall's Ledge in September 1955. HMS *Vidal* stands off in the background.
(Natural History Museum)

Left: It's got a Union Jack flying on it – so it must be British! First Lieutenant Scott from HMS *Vidal* hoists the flag in September 1955 watched by Corporal Fraser
(Natural History Museum)

algae covered rock was difficult enough, even when the reef was comparatively flat and level, but Hall was attempting to leap onto a rock that appeared to rise vertically out of the Atlantic. His narrative continued;

One side of the rock was perpendicular and as smooth as a wall. The others, though steep and slippery, were sufficiently varied in their surface to admit of our crawling up when once out of the boat; but it required no small confidence in our footing, and a dash of that kind of faith which carries a hunter over a five-bar gate to render a leap at all secure. In time, however, we all got up, hammers, sketchbooks and chronometers inclusive. As it was a point of some moment to determine not only the position but the size of the rock by actual observations made upon it, all hands were made busily to work – some to chip off specimens, others to measure the girth by means of a cord – while one of the boats was sent to measure soundings in those directions where the bottom could be reached.

They busied themselves with their work and were at first unaware that the visibility was becoming distinctly hazy. This thickened into a fog and blotted out any sign of their ship. Again, we must compare events on Rockall with those earlier in this volume on the Bell Rock in September 1807 when Stevenson and his men were stranded on the rock as their tender, the *Smeaton*, drifted away after its mooring ropes were snapped by worsening weather. He was lucky, and all were rescued by an Abroath pilot boat. Now, just three years later, Basil Hall and his men were marooned on the most isolated rock in the Atlantic Ocean with a fog descending and their own ship nowhere to be seen.

The swell had silently increased in the interval to such a height that the operation of returning to the boats was rendered twice as difficult as that of disembarking, and, what was a great deal worse, occupied twice as much time. It required the greater part of half and hour to tumble the whole party back again. This proceeding, difficult at any season I should suppose, was now reduced to a sort of somersault or flying leap; for the adventurer whose turn it was to spring, had to dash off the rock towards the boat trusting more to the chance of being caught by his companions than to any skill of his own.

From the open boat there was still no sign of *Endymion*, so someone suggested that a single lookout posted on top of Rockall would have a better view 70 feet higher up. One of the crew was offloaded back on to Rockall and "skipped up the rock like a goat". The only thing he saw was another fog bank approaching which reduced visibility to less than ten yards. After a period of time, in which the inhabitants of the boat drifted in the lee of Rockall and considered their fate, their was a sudden cry from the watchman, "I see a ship!" The mists had parted sufficiently for a ship to be glanced through an opening in the fog. It took another 15 minutes to retrieve the lookout before the whole crew rowed strongly towards their ship as the mists closed again, blotting out Rockall behind them.

Then, disaster! The *Endymion* suddenly tacked away from them – obviously not having spotted the lost souls – straight into another fog bank. Their ship had disappeared again, and they were some distance from Rockall. They made the only sensible decision available to them – to row at top speed back towards the shelter of the rock again to wait until, hopefully, conditions changed. During the wait they discussed matters like what to do if the weather worsened and they had to abandon their boats – like hauling one of them to the summit of Rockall, inverting it and using it as a kind of shelter!

The sun set, and just as the hapless crew were turning their thoughts to abandoning their boats for the comparative stability of Rockall again, the fog lifted to reveal the *Endymion* standing a short distance off Rockall – unaware of their exact position. It was almost dark when the crew regained their ship, bringing to an end the day that Rockall was first, knowingly, conquered. Hall's legacy of that day was the naming of the only ledge on the whole rock where it was possible to stand on level ground, 13 feet below its summit, 'Hall's Ledge' – a name it still carries to this day.

Incidentally, Hall's survey of Rockall's position was surprisingly accurate given the instruments he used in 1811. His calculations put it at 57° 39′ 32″N, 13° 31′ 16″W which is only about seven miles north-east of the accepted Admiralty chart position based on data gathered in 1831. Also, note the date of Hall's ascent of Rockall – 8th September 1811. Most accounts of this historic first landing put it on 8th July 1810 – an error perpetuated in most subsequent accounts. Thanks to some clever detective work in the ship's log by James Fisher, an ornithologist who has landed on Rockall and wrote a book about it in 1956, the July 1810 date is clearly incorrect because on that date the *Endymion* was sailing within sight of Smeaton's Eddystone lighthouse.

In 1831 the Royal Navy were back in the waters around Rockall for further surveys. HMS *Pike* under Lieutenant Commander John Wigley spent 6th–20th September on the Rockall Bank. On board was Captain Alexander Vidal, one of the navy's greatest surveyors who fixed Rockall's position at 57° 36′ 20″N, 13°41′ 32″W.

Some 31 years later, another Royal Navy visit to Rockall with HMS *Porcupine*, under the command of Captain Richard Hoskyn, occupied June and August of 1862. They were undertaking survey work for the proposed laying of an Atlantic cable across the underwater reef of which Rockall is the summit. On 15th August 1862 they managed, with some difficulty, to get one man – Boatswain Johns – on to the rock, but he was unable to climb its precipitous sides before being taken off. He did managed to collect a couple of rock samples during his brief stay, but more importantly, he had made only the second authenticated landing on Rockall. The same vessel returned for a further survey in 1869.

In 1888 there is a report of a skipper from a Grimsby fishing smack landing on Rockall and making the summit while in 1948 one of the crew from the Fleetwood trawler *Bulby* reportedly swam around the rock with the aid of a float! A party of brave Irishmen attempted an assault on Rockall in 1896 – Robert Lloyd Praeger and colleagues set out from Killybegs in

South-west Donegal on board the steamer *Granuaile* to survey the rock on behalf of the Royal Irish Academy – but only managed to make water colour sketches and two abortive landing attempts before taking shelter at St Kilda, exhausted by two rough weeks rolling at sea. But they did manage to take the first ever photographs of Rockall.

The results of HMS *Porpcupine*'s two surveys have given us much of the information we now have on Rockall. The Rockall 'Bank', as it became known, has a length of about 100 miles and a width of 60 miles, most of which is covered by only 60 to 100 fathoms of water – very shallow compared to the virtually bottomless depths of the Atlantic on either side of the Bank. Closer to the rock it's even shallower – a mere 20 to 30 fathoms – and in two places there are actually outliers to Rockall itself. One cable (approx 608 feet) north of Rockall, and separated from it by a channel with 30 fathoms in it, is Hasselwood (or sometimes 'Haselwood') Rock – a small drying pinnacle of granite, but much smaller than Rockall itself.

Two miles ENE of the rock lies Helen's Reef, which rises barely five feet out of the water and was named after the wreck of the *Helen*, a brigantine out of Dundee. On 19th April 1824 she was bound for Canada with a general cargo when, in broad daylight and unsure of her position, she struck twice on the reef to which she was subsequently to give her name. All the crew, numbering twelve, and one passenger made it into two lifeboats, while the remainder of the passengers – three men, seven women and six children – drowned when the *Helen* eventually sank thirteen hours later.

The Admiralty Pilot for the area describes a pretty unforgiving environment for vessels passing close to Rockall. Part of it reads:

5 miles E of the islet the sea has been reported to break. In addition, a large number of pinnacles, with depths of less than 100m over them, lie within 6 miles of Rockall. The summit of this reef has been seen in the trough of the sea at LW spring tides, but towards HW and in fine weather, the sea only breaks upon it at long intervals. There is nothing except the breakers to indicate approach to this reef...

Imagine what a frightening experience this would have been for early mariners in the days before all these dangers were charted. Approaching Rockall in heavy seas and at low water, and just about to pass over the crest of a large wave – only to see in the trough in front of them the teeth of an unmarked reef appearing out of the swirling foam. In darkness, of course, tragedy was inevitable with little chance of rescue for any survivors 300 miles out into the Atlantic.

Exactly how many vessels have been lost in the vicinity of Rockall we shall probably never know. However, we do know about the Danish liner S.S. *Norge* that was crowded with emigrants making their way from Copenhagen to New York for a new life in America in 1904. On 28th June the weather was sufficiently bad to drive her off course causing her to strike Helen's Reef in haze. She sank in 20 minutes with the loss of 582 passengers and 45 crew. The few lucky survivors who made it into one of the five serviceable lifeboats were eventually rescued by passing vessels and landed in such diverse locations as Stornoway, Grimsby, and Torshavn in the Faroe Isles. These tragic events took place only 8 years before the *Titanic* would strike her iceberg with similar consequences.

Such a tragic event only heightened public awareness of an unlit rock off the coast of Scotland. Not for the first time the Commissioners for Northern Lighthouses were asked about the possibility of putting a light on Rockall. Their chief engineer at the time was David Alan Stevenson. He had never set eyes on Rockall but, having talked with a few people that had and by reading what little detail existed in print about the rock, he was nevertheless of

Above: Rockall's pinnacle has been removed to give a flat area to fix the mounting ring for its first beacon in June 1972. A Sea King helicopter from 816 Squadron hovers above.
(Imperial War Museum)

an opinion that, "...while it would not be impossible, though very costly, to erect a lighthouse station on Rockall the difficulty, or rather the impossibility, of relieving it with anything approaching regularity, makes it a scheme that cannot be entertained."

Stevenson did suggest, however, that some kind of lightship moored close to the rock was a possibility. This would not be a traditional lightship – a vessel devoid of engines, towed to her correct station and then moored with huge lengths of massive chains – but one with engines powerful enough to steam out to Rockall and keep her on station during winter gales. In fact, he proposed two such vessels so one could relieve the other, and suggested they be fitted with "wireless telegraph apparatus" which would be "very valuable from a meteorological and storm, or weather forecasting point of view." Nothing further was heard of Stevenson's idea.

Part of the research for this chapter involved studying a curious little publication – a 33 page pamphlet called *Rockall* by James A. Macintosh. It was published in 1946 and contains all sorts of interesting and quirky information about Rockall. It's an entertaining mix of fact, conjecture and opinion about Rockall and other topics dear to his heart such as tidal power. Some of the opinions and proposals put forward by Mr Macintosh are – with the benefit of hindsight – strangely prophetic, others are just flights of fancy. For instance, writing about the wreck of the *Norge* he says:

To avoid a recurrence of such a tragedy as happened to the "Norge" I would suggest that Rockall be sprayed with white paint so that it could be spotted easily in daytime, and that a bright light should be placed upon the summit during darkness to warn mariners on board ships at sea, and give pilots of air-ships in the sky their exact position. As an afterthought I would suggest coloured lights at the four corners of the square to mark the four points of the compass – red for North, green for South, blue for East, and yellow for West, with the white central beam of light reaching up to the stratosphere (and don't forget the white paint).

No mention, of course, as to what should happen when the first winter storm strips the paint – and all the coloured lights – off the rock, but his idea for a light on the summit was enhanced by an elaborate description for a structure constructed from metal hoops and bands to encase the entire rock. Read for yourself, the section of his pamphlet that details this idea.

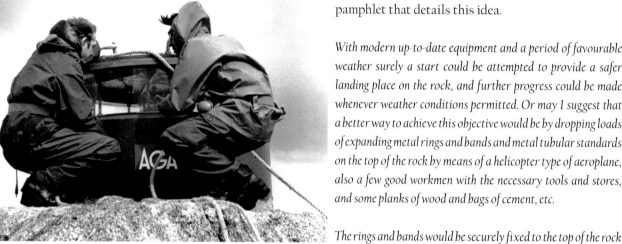

With modern up-to-date equipment and a period of favourable weather surely a start could be attempted to provide a safer landing place on the rock, and further progress could be made whenever weather conditions permitted. Or may I suggest that a better way to achieve this objective would be by dropping loads of expanding metal rings and bands and metal tubular standards on the top of the rock by means of a helicopter type of aeroplane, also a few good workmen with the necessary tools and stores, and some planks of wood and bags of cement, etc.

The rings and bands would be securely fixed to the top of the rock and from there continued downwards by stages. All chips of rock would be collected with care and the deposit of guano on the top put into bags, and the top of the rock levelled out for the landing of further supplies by helicopter. A cylindrical saloon hut for the accommodation of the workmen cold be suspended over the side of the rock from the top circular rings so that the hut could be moved as required around the rock and thus be always on the sheltered side. Other huts would be required for stores, engine, etc.

When the tubular scaffolding reaches down to sea level, my idea is to build a circular wall round the rock of metal interlocking blocks of an aluminium alloy, about 250 feet in circumference and 70 feet or thereby upwards to the level of the summit.

The tower would be built solid from the base of the rock to about halfway up, and then compartments would provide accommodation for all purposes, and the outside huts would no longer be required. The metal blocks, being deposited by plane on the summit and being of a standard size, would only need lowering into position and fixing. When this stage is reached the rock would have completely changed its appearance, and would now look something like a huge white Dunlop cheese floating on the sea. The birds would have all gone - flitted to some other sanctuary. Some say that the wild Atlantic waves could sweep clean over the rock, but I doubt it, in any case the building upwards to any desired height could easily be continued.

His engineering naivety appears to be breathtaking. His talk of building with planks of wood and bags of cement 300 miles into the open Atlantic, or dangling workmen's huts over the side of the rock from circular rings, or collecting all the rock and guano in bags to make a helicopter landing pad, makes us smile. But there is a subsequent paragraph that makes you wonder if Mr Macintosh was a visionary who had more foresight than we credit him with.

Bear in mind that the words below were written in 1946:

I now appeal for an up-to-date Radiolocation station also. More accurate and more frequent weather forecasts are a vital necessity. Rockall would provide an ideal spot, and if Trinity House still will not undertake the job then St. Andrew's House must. The advent of Radiolocation may in the course of time supersede the lighthouse to a certain extent, but even that new invention will not be immune from accidental breakdowns, and so for additional safety to land, sea and air transport. of the future, every safety device will be necessary. The elimination of the guiding lights of the lighthouses, and beacons, and warning notes of the foghorns, should never be sanctioned.

Interestingly he mentions Trinity House, even though Rockall would officially be in Scottish waters and therefore any lighthouse would be the responsibility of the Commissioners of Northern Lighthouses. This was because Captain Hall from HMS *Endymion* – fresh from his first documented landing on Rockall in 1810 – made a subsequent appeal to Trinity House for some kind of warning device to be placed there, either a "beacon light" or a lightship. Hall also acknowledged the difficulties of such an undertaking, not just the landing of building supplies but also "the difficulty of procuring any individuals to man the lighthouse on a rock of such extreme isolation." Trinity House, with lights already functioning on the Eddystone, Smalls, Bishop Rock, and Longships, would be well aware of those difficulties.

But the readily acknowledged difficulty of building, manning and relieving the keepers of a lighthouse on Rockall wasn't the real problem. The real problem was that Rockall wasn't actually part of Britain. Although the nearest mainland to Rockall was Scotland, and the Royal Navy had surveyed it on several occasions, it had never been formally annexed to the United Kingdom. Therefore Trinity House, or for that matter the Northern Lighthouse Board, couldn't officially erect a lighthouse on a piece of rock that wasn't part of England, Scotland, or Wales. Nobody owned Rockall. And this is where the story gets really complicated – because politicians now become involved!

It was used during World War II by the RAF in navigational exercises, and it was also planned to be a 'target' for a long range missile programme during the bleak years of the Cold War in the 1950s. The idea was that the UK's first guided nuclear missiles would be fired at Rockall from a secret base on South Uist in the Outer Hebrides to test their accuracy. This plan caused the British Government to then classify Rockall as a "security risk"! Whitehall feared that the Soviet Union might occupy the rock and build an observation

Left: A government publicity photo taken shortly after the first beacon was put on Rockall in 1972. It shows a couple of marines in full dress uniform guarding a sentry box flying the white ensign flag! Rockall was definitely British now!
(Crown Copyright)

platform or place measuring instruments on the rock in order to track missiles being test-fired at it – because, officially, nobody owned Rockall. The Foreign Office stated "the immediate use for Rockall is nebulous, but it is hoped that in time it may provide an aiming target, and it would be rather unfortunate if someone claimed it first and used it as an observation point".

This was serious stuff and required decisive action. In April 1955 a message was sent to the Admiralty to 'seize' Rockall and declare its sovereignty to be British. Consequently, Rockall was annexed by Britain on 18th September 1955 by a landing party from HMS *Vidal*, a 2,000 ton survey ship. A helicopter flew the party onto the rock because, at that time, the *Vidal* was the only helicopter-carrying research vessel in the world. Foreign Office lawyers advised that "some physical sign of the claim must be left behind" so the party cemented a brass plaque to the rock, hammered in climbing spikes and added ring bolts near the water line before hoisting the Union Jack. Part of the plaque read, "…in accordance with Her Majesty's instructions dated the 14th day of September, 1955, a landing was effected this day upon this island of Rockall from HMS *Vidal*. The Union flag was hoisted and possession of the island was taken in the name of Her Majesty." Lieutenant Commander Desmond Scott claimed it for the UK with the words "In the name of Her Majesty Queen Elizabeth II, I hereby take possession of the island of Rockall". He then returned to his boat and fired a 21-gun salute.

On 22nd September 1955 *The Times* reported the Admiralty had announced "the formal annexation of Rockall" for the Queen. A Ministry of Defence spokesman said, "Rockall belonged to no nation. It had been formally claimed by the Crown to eliminate the possibility of embarrassing counter-claims once the Hebridean guided missiles project was underway". Finally, the vexed question of Rockall's nationality was assured. Well, not quite! According to the UK Government the annexation made Rockall part of the "dominions" of the Crown, but it did not make it part of the United Kingdom.

The Royal Navy's visit to claim Rockall for the Crown also produced a counter claim by the Clan Mackay. Apparently, an 84 year old Highlander Councillor, J. Abrach Mackay, announced "My old father, God rest his soul, claimed that island for the Clan of Mackay in 1846 and I now demand that the Admiralty hand it back. It's no' theirs." Another member of the committee commented to the newspaper, "It wouldn't surprise me to see a battleship appearing off Rockall any day. You can't afford to underestimate the Mackays". Apart from this sabre-rattling, there was little else in the way of documented proof, and so no more was heard from Councillor Mackay.

Four years later the Royal Navy were back at Rockall with HMS *Cavendish*, only to find that the brass plaque from 1955 had completely disappeared – so they left a stone tablet instead and inscribed the name of their ship in some quick drying cement. In 1969 they were back again. This time they arrived by inflatable boat on 25th March and unfurled a White Ensign on the summit. The *Daily Mail* reported on 17th April that, "it's absolutely and undeniably British. The Navy has just invaded it for the third time – to make quite sure". The sailors then "posed for photographs…lasting proof that, to use Shakespeare's words, this fortress built by nature for herself is still as British as the River Thames".

In the same year an intensive programme of geological research in the area was initiated by the Government in the search for oil and gas. It was undertaken by the Natural Environmental Research Council through the National Institute of Oceanography and a team from Cambridge University. The Government also confirmed that it would "assert for Great Britain the exclusive right to exploit resources on or under the Rockall Bank". Oil? Gas? And these were in addition to Rockall's already confirmed bounteous fish stocks. Suddenly, almost overnight, Rockall became a very desirable piece of real estate.

On 9th October 1970 *The Guardian* reported that their findings revealed they may have indeed found something a little more valuable than fish breeding grounds. "Rockall may yield gas. There may be oil or natural gas on the Rockall Plateau in the North Atlantic 300 miles north west of Ireland. The research ship *Discovery* has reported a new sedimentary basin on the plateau. Similar sedimentary basins exist in the Irish Sea, where gas has already been found."

Oh dear. The prospect of untold wealth under the seas around Rockall started a bit of a rumble in Whitehall. Don't forget, Rockall was only "part of the dominions of the Crown" and not officially part of the United Kingdom. Not only that, but *The Guardian* had found some of the geologists who took part in the 1969 survey and they were quoted as saying that, geologically, Rockall was more likely to be part of North America and Greenland than Scotland!

"It cannot therefore, really be said to form part of the Continental Shelf of Great Britain". This was all extremely relevant, the paper said, "...because of Britain's adherence to the International Convention on the Continental Shelf, which lays down that we can dig up anything from oil to oysters on any bit of sea around us which has less than 200 metres of water over it. We are allowed to go beyond that if the depth allows exploitation of natural resources and in areas adjacent to the coasts of islands. "The snag is" said *The Guardian*, "that between oil-rich Rockall (as the guide books will doubtless say) and Scotland, there is a hole in the sea bed about 400 miles long, 80 miles wide and 9,000 feet deep. It is not...what you might call pipeline country... but the whole area is, of course, "adjacent to Rockall". Does an uninhabited island come within the terms of the Convention? It seems unlikely. Can it be argued that Rockall is an integral part of the United Kingdom? That seems pretty doubtful".

Left: A couple of technicians can be seen on Hall's Ledge after their maintenance visit to the second of Rockall's beacons.
(Pharos Marine)

Things were not looking at all shipshape for HM Government. Conveniently ignoring the geological evidence produced by their own experts, they trotted out the argument that because it was part of 'the dominions of the Crown', and that our Royal Navy had landed on it, surveyed it, erected the Union Jack on it, and attached brass plaques and stone tablets to it on more than one occasion it was rightfully ours, and we were keeping it. This didn't seem to worry countries like Denmark (whose Faeroe Islands were only 400 miles from Rockall), Ireland and Greenland who all of a sudden were taking a tremendous interest in the sovereignty of Rockall and started producing maps showing which parts of the sea around Rockall they owned!

Behind the scenes things were obviously happening a furious pace. In 1971 the Royal Navy and some more geologists were back on Rockall. The *Sunday Telegraph* reported, "...the purpose was threefold; to seek indications of natural gas or oil, to prepare a site for a navigation light and possibly a helicopter landing platform, and to reinforce Britain's claim to the rock." There was even mention that our old Cold War adversaries, the Russians, might try to land and annex Rockall.

To give all this some air of legitimacy, on 4th November 1971 the *Isle of Rockall Bill* was presented to the House of Lords by Lord Campbell of Croy. The Queen gave her Royal Assent to *The Isle of Rockall Act* on 10th February 1972 since when it has been part of Invernesshire. Part of the debate produced two interesting quotations from those involved. Firstly, Lord Kennet offered his opinion that, "There can be no place more desolate, despairing and awful." and few would disagree with that, and from William Ross, MP for Kilmarnock, the fascinating revelation that, "More people have landed on the moon than have landed on Rockall."

At last, the waters around Rockall were becoming clearer. The possibility of huge

Above: A spectacular view of Rockall in June 1997 when Greenpeace activists occupied the rock. They were protesting at the continuing exploration for oil in the deep waters of the 'Atlantic Frontier'. They lived in the yellow 'survival pod' that can be seen strapped to Hall's Ledge. The beacon on the summit is clearly different to the first two placed there.
(Greenpeace - Sims)

reserves of oil and natural gas under the Rockall Bank prompted the Government to officially annex the rock to the United Kingdom before another country annexed it for themselves. And just to make sure that everyone knew it belonged to the United Kingdom, we were going to put a navigation beacon on the top of it. The *Daily Mail* summed it all up very succinctly, "Too small even to be a dot on a map, its only claim to fame is its distinction of being further from the mainland than any other rock of its size in the world. It now seems a safe bet that Rockall will no longer be left to the gulls, guillemots, kittiwakes and the weather men".

In June 1972 the Government despatched "a combined expedition" to "set up a flashing navigational beacon" to aid shipping. Finally Rockall was to get its 'lighthouse'. But before this could be fitted there was some formidable preparation to be done on the rock. Nowhere on Rockall was there an area of level rock to place a beacon that would be visible from every angle – and that included the summit which was a pinnacle of granite. It was decided to remove this and 'decapitate' the rock leaving a flat surface approximately 12 feet by 3 feet – just enough room to fit a 'lantern' containing a phosphor bronze flashing beacon supplied by the AGA company. By drilling a series of transverse holes right through summit of the rock, a party of Royal Engineers removed sufficient granite to give Rockall its remodelled profile before the installation could start in earnest.

This time, it was a 15 day operation involving engineers from the Department of Trade and Industry, scientists and divers. They were taken to Rockall, together with two Sea King helicopters from 816 Squadron, on board RFA *Engadine* that left Plymouth on 16th June. *The Times* even had readers writing letters to ask if the construction of the beacon would affect the birds on the rock. One correspondent wrote about the force of the prevailing wind and swell being "…truly remarkable. During the recent visit of RFA *Engadine*, equipment competently secured on or near the ledge with pitons and ropes of breaking strains of thousands of pounds was carried away virtually without trace." He writes with such authority and detail that it's probable he was actually part of the expedition – or had detailed knowledge of it. But, how many times in this volume have we seen instances of the ocean's power? Rockall is no different to any other wave-swept rock in British waters, in fact it's probably ten times worse, and the force of a sea in spate should no longer surprise readers who have come this far.

Two teams of men were at work over the 15 days the rock was occupied. Above the waves the engineers and technicians laboured to install the lantern on a concrete base and moved around the rock attached to safety ropes previously attached by two Royal Marines. Often, it was a case of snatching brief periods of work in between the sudden squalls and gales – just like Smeaton, Whiteside, Stevenson and others had done a century before. Below the waves a team of divers surveyed the rock, Helen's reef, the sea bed around them, and also collected samples they hoped might gives clues as to their origins. Samples were also taken from the rock itself by geologists from the Institute of Geological Sciences. Before the RFA *Engadine* sailed, another plaque was fixed there for good measure.

The navigation light was designed to run without maintenance for a year at a time and would be the responsibility of the Government rather than Trinity House or the Commissioners for Northern Lighthouses. According to the Admiralty List of Lights, the beacon on Rockall exhibits a single white flash every 15 seconds from a red lantern that is visible for 8 miles. However it does issue a warning about its reliability – which is not particularly good. Winter storms around Rockall can be so severe that the beacon is frequently unserviceable for long periods and the rock is returned to darkness for weeks, and sometimes months at a time.

A study of the small number of photographs that actually show Rockall's beacon

Left: A dramatic demonstration of why the beacons placed on Rockall don't seem to last. By comparing the height of this massive wave – captured by the RAF in March 1943 – with the height of the rock it can be estimated that the height of the column of water is about 170ft, or over 2½ times the height of Rockall. (*Natural History Museum*)

reveals that several different designs have been placed on it. The original 1972 AGA model looked very much like a traditional lighthouse lantern with the flashing bulb housed inside a circular lantern 'room', about 3 feet high, complete with glazed windows and conical roof. Clearly this beacon didn't survive because subsequent photographs show a much-reduced structure – less glazing and more reinforcement around the lantern, which was now mounted on a substantial circular steel drum type of structure that housed the batteries. Even later photographs show a beacon with a more 'streamlined', almost triangular, profile – the better to resist the damaging effects of huge waves perhaps? However, dating all these different design changes has proved next to impossible.

While all the various beacons on Rockall were being installed or modified, the political

Right: In July 1997 Greenpeace activists ended their 48 day occupation of Rockall. Here we see them lowering the flags of Great Britain, Denmark, Iceland, Ireland and the Faroes as the Waveland flag is raised. During their stay they converted the beacon on Rockall to solar power, the panels for which can be seen attached to the back of Halls Ledge.
(Greenpeace - Sims)

Left: The following year, in 1998, Greenpeace installed another solar powered beacon to replace the original one as part of their Atlantic Frontier campaign. The light, although slightly modified, was reported to be still functioning six months later.
(John Cunningham)

ramifications of Rockall's annexation continued to rumble on many hundreds of miles away from the rock itself. In December 1982 the United Nations Convention on the Law of the Sea was opened for signature by interested nations. This is the convention that set the 12-mile territorial waters limit and, more importantly for Rockall, a 200-mile Economic Exclusion Zone (EEZ). The baseline for the start of an EEZ is a line drawn between the outermost points of the outermost islands. At this time of course the British government were claiming that Rockall was one of their outermost islands, which therefore meant their EEZ – containing all the potential oil and gas reserves – extended for a further 200 miles to the west of Rockall.

One of the stumbling blocks for its ratification by the United Kingdom was a clause that stated, "Rocks which cannot sustain human habitation or economic life of their own shall have no exclusive economic zone or continental shelf". Clearly Rockall could not sustain human inhabitation or economic life and so reduced the UK's EEZ considerably, but, as Rockall does lie within 200 nautical miles

Left: The remains of the Rockall beacon in 2005. The 1998 solar powered light obviously met the same fate as its predecessors and has been swept into oblivion. However, the plinth has provided a very convenient and sheltered nesting site for a pair of guillemots.
(John Cunningham)

of St Kilda and North Uist, the island itself remains within the EEZ of the United Kingdom. Furthermore, the United Kingdom and Ireland also signed a boundary agreement that includes Rockall in the United Kingdom area.

However, there were some intrepid individuals who challenged the assumption that it was impossible to sustain human inhabitation on Rockall. In 1985 a former SAS soldier, survival expert, and lone Atlantic sailor called Tom McLean arrived at Rockall in a fishing boat and lived on the rock from 26th May to 4th July in a wooden shed bolted to Halls Ledge! With that

particular feat of endurance, Tom had shown that the island was not only part of the United Kingdom, but also capable of sustaining human habitation, thereby re-affirming Britain's mineral and oil rights to the rock.

On 10th June 1997 four members of the environmentalist organisation Greenpeace were winched onto Rockall from a helicopter and occupied the islet for a time. They declared it to be a "new Global State" called Waveland – which existed only on the internet – as a protest against oil exploration under the authority of the British and offered citizenship to anyone willing to take their pledge of allegiance. The British Government's response was simply to give them permission to be there, and otherwise ignore them. Part of the Waveland web site explains; "Waveland is a new global country situated on the internet. Its citizens share common values: co-operation, respect for difference, peaceful co-existence with one another and with nature. It exists to allow its global citizens to meet and imagine how things might be otherwise." For 48 days the Waveland settlers lived in a plastic solar-powered survival 'pod' on Hall's Ledge – the longest human inhabitation of Rockall yet. One of their philanthropic deeds was to erect a solar powered navigation beacon on the summit – to replace the official one that was unserviceable due to storm damage.

At the time of writing the most recent landing on Rockall was on 16th June 2005 when a team of radio hams landed for a brief few hours before approaching weather fronts forced them to abandon their occupation.

Rockall's beacon continues to flash its warning – winter storms permitting – in what must be the most hostile environment ever for such a device. Born out of political expediency in order to extend the boundaries of oil exploration for Britain, it nevertheless manages at the same time to fulfil a humanitarian role. Rockall is a wild, god-forsaken lump of rock that, I imagine, fewer than a hundred people have ever set foot on. Most of these people will have shared the summit with the most isolated lighthouse in the world.

Left: The Saltire flies from a 30ft radio mast erected on Rockall by radio hams Seamus Cameron and David Wood as they "activated" the rock for the first time on 16 June 2005.
(John Cunningham)

16 ROCK LIGHTS

a new era begins

THERE CAN be no doubt that there will never be another rock light constructed around the British coastline of a similar design to such legendary giants as Bishop Rock, Eddystone, Bell Rock or Skerryvore. The days when men like the Stevensons, the Douglass's or James Walker could make their reputation and fortune by designing and building lighthouses on some of the most exposed sites in the world have gone. The few lighthouses that have been constructed since the middle of this century are the product of relatively anonymous engineers, built of modern materials to modern designs, and full of technology that was designed to make the job of a lightkeeper unnecessary.

The rock lighthouses that appear in this book mark a particular era in civil engineering achievement that has now passed. The newest rock lighthouses, which in themselves are rare phenomena, are designed to be totally automatic in operation and functional in design – no longer is a new lighthouse considered a thing of grace and beauty. It is also sad to relate that even the old masonry towers have changed as part of their conversion to automatic working. For some of the towers the first feature to go was their graceful symmetry – changed forever by the addition of a helideck above the lantern which shrouded the sparkling glassware inside the lantern behind a latticework of steel. This was followed by the less obvious changes to internal equipment and illuminating devices.

The 'Golden Age' of British lighthouse construction spanned the century between roughly 1790 and 1890. This was a period when most of the worst reefs, promontories and islands received a lighthouse, nearly all of which survive to this present day. Prior to this time was the period of innovation when engineers were tentatively placing their structures of various designs and efficiency upon rocks previously thought an impossible site for any kind of permanent construction. Henry Winstanley had put two towers on the Eddystone followed by Rudyerd's wooden light before John Smeaton finally led the way for modern lighthouse design with his cylindrical stone pillar. Let us not forget also that Henry Whiteside had placed a wooden, open-piled lighthouse on the Smalls before 1790, in a design that has been copied throughout the world in certain situations.

SMEATON'S CHANDELIER, 1759.

Left: Early illuminants for rock lighthouses were usually tallow candles carried in a chandelier. This illustration shows the 24 candles that were used in Smeaton's Eddystone of 1759. *(The Strand Magazine)*

All of our present acclaimed rock towers are of post-1790 construction. Samuel Wyatt's short, stumpy Longships tower marked the start of the great renaissance of lighthouse construction in 1795, followed by a remodelled Skerries light in 1804, Bell Rock in 1811, Longstone in 1826, Skerryvore in 1844, and the first stone tower on Bishop Rock in 1858. In quick succession these were followed by stone towers on the Smalls (1861), Wolf Rock (1870), Dubh Artach (1872), the second Longships tower (1873), Chicken Rock (1874), and the present towers of Eddystone (1882) and Bishop Rock (1887). Nearly 100 years of some of the finest civil engineering to be seen, yet so often never fully appreciated because of its remoteness. These were the years of Nicholas, James and William Douglass, of Robert, Alan, David and Thomas Stevenson, Joseph Nelson and James Walker – all brave and talented men who fought with nature, and won. They produced structures that are every bit as functional and every bit as sound today as when they were built, despite a constant assault by over a century of storms and revolutionary advances in technology. In the age of the micro chip we have yet to devise an adequate replacement for the lighthouse.

The period between the turn of the twentieth century until about the late 1960s/ early 1970s saw progressive changes to the life and routine of a lighthouse keeper. While

Opposite: The Trinity House helicopter hovers dramatically alongside Longships lighthouse during a storm. On its helideck is a rubber water bag that has been delivered by the helicopter. *(MBB)*

the structure in which he lived changed outwardly very little since it was built, the duties of a lightkeeper, the equipment in his charge, and his general day-to-day routine underwent considerable changes as the decades progressed. Originally, life for the keeper of a rock station

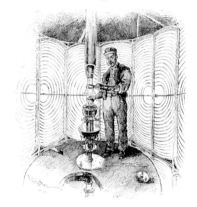

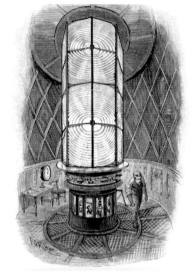

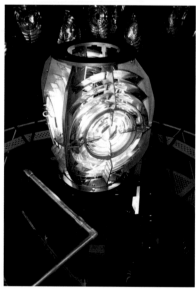

was indeed a lonely and remote vigil in an outpost of the British Isles. Confined to a stone tower with two or three other men, out of contact with his family and the mainland for weeks at a time, and only relieved when the weather was sufficiently calm to allow an open boat close to the tower. His only source of power would have been paraffin or colza oil, possibly coal for cooking, and his entertainment when off duty was purely what he brought with him or could devise himself. During the winter months a turn of duty could be prolonged by many weeks by the weather, with all the attendant boredom and frustration that meant.

By the end of the 1970s such an existence had, by and large, disappeared. Since then, the day-to-day existence for a keeper at a rock station was pretty well as it would be at a shore station, apart for the absence of his family. Huge diesel oil generators provided the electricity for power and light, including the high-wattage electric bulbs in the lantern that replaced the paraffin and acetylene burners. Electricity within the tower meant that television and radio, plus the convenience of modern appliances such as fridges and freezers were all available to the lightkeeper, as well as instant communication with the mainland by VHF radio and UHF telephone if necessary. Internal fitments and fittings were all upgraded so that life in a modern British rock station was altogether a more civilised existence than it used to be.

But, there was one particular aspect of life in a rock light that changed out of all recognition within the final years of manned operation – the method of relieving the keepers. One of the greatest hardships of a posting to a rock station was the uncertainty of relief, for it was only on a near perfect day that such a hazardous manoeuvre could be attempted at many of these stations. In most cases it involved navigating an open boat close enough to the base of the tower to allow a rope to be stretched between them. All men and supplies were then exchanged by being dangled from a pulley that ran along this rope. If the rope lost its taughtness, then whatever was travelling along it was in danger of being dumped into the confused waters below the tower. It was, by all accounts, an extremely nerve racking procedure for both those in the boat, who had to keep the craft at a constant distance from the tower in a not inconsiderable swell, and for the keepers who were being changed, although many regarded it as part and parcel of their life at a rock station. Some rock stations were on sufficiently large reefs to allow a boat to tie up at a jetty that had been constructed specifically for this purpose, in which case a relief became considerably less fraught.

Clearly, a situation where reliefs could be delayed for anything up to six weeks in the winter because of the sea was most unsatisfactory and could throw the relief schedules into chaos. In the early 1970s Trinity House and the Northern Lighthouse Board solved this problem by effecting reliefs by helicopter. Both Trinity House and the NLB hired Bolkow twin-engined helicopters which were used to relieve their remaining manned stations every 14 or 28 days, or deliver maintenance and engineering personnel to the unmanned towers. The helicopters set down, either on an area of reef adjacent to the tower, such as is the case at Longstone, Dubh Artach and Skerryvore, or on a specially constructed helideck, 30 ft in diameter, held above the lantern by an elaborate web of steel supports. As I have mentioned on more than one occasion, this 'mortar board' headgear does little to improve the appearance of the tower – many would argue the opposite – yet there can be little argument that this viewpoint is offset by the amount of time and expense saved in relieving the keepers on time or with only minimal delay. The first British lighthouse to have its keepers relieved by helicopter was Wolf Rock from its newly constructed helideck on 3rd November 1973, although a helicopter had been used to deliver overdue supplies to this station as early as 1948.

The advantages of such an operating procedure were enormous, and the conversion of all the major rock stations to facilitate helicopter reliefs was tackled with some urgency. All the famous English towers – Wolf Rock, Bishop Rock, Longships, the Smalls and Eddystone – were given helidecks in succession. Interestingly, there is no Scottish rock light with a helideck above its lantern, as all the Northern Lighthouse Board towers have room for a helicopter to set down beside the tower.

Below: A Bishop Rock boat relief has been captured in this old postcard that shows how the keepers were transferred between an open boat and the set-off of the lighthouse. Most of the other boats in attendance are filled with tourists who came to watch the spectacle.

At the Bell Rock, where the strata of the rock made it impossible, a special raised helipad was built on the reef itself to enable a helicopter visit that coincided with a low tide during daylight hours.

Up to three keepers could be exchanged on each relief, but only one or two was the usual practice, as well as about 270 kg of supplies and provisions. Reliefs were rarely delayed by more than 24 hours because of the weather, fog being the main culprit in this respect, and they have still continued in force ten winds. Each helicopter has instrumentation to allow it to fly in poor visibility or at night. The keepers were flown from one of the land bases around the British Isles where their homes were normally situated. In addition to carrying crew, the helicopters are also used to carry personnel from the engineering departments to and from the lighthouses with equipment and materials. An underslung cargo hook is fitted to each helicopter. In conjunction with this, the two Trinity House vessels – THV

This page: A sequence of photographs showing a typical Wolf Rock boat relief in the 1920s before it had a helideck. Communication with the lighthouse as to whether a relief was going to be attempted was all done by semaphore flags from the Trinity House tender. *(Syndication International)*

Right: The transfer from open boat onto the landing stage was done by a metal derrick pole which can be seen in action here. On the very left is the small boat waiting to receive or discharge the keepers.

Left: Mr F. G. Few seems to be taking his relief from Wolf Rock in March 1923 all in his stride. He has dressed in overcoat and bowler hat for the occasion and still smokes a cigarette as he is lowered into a waiting boat, clutching what is probably his fishing rod. *(Syndication International)*

Far left: A view of the transfer procedure from the tower itself. Boiler ash is spread on the landing stage to enable the workers to keep their footing on what could be a trecherous surface. *(Syndication International)*

Left: A ship's officer from the Trinity House tender is brought onto Wolf Rock in Septmber 1972. The triangular metal beacon built into the landing stage is on the left of the picture carrying a Wolf Rock lifebuoy. *(ALK archives)*

Patricia and THV *Mermaid* – as well as the NLB tenders MV *Pharos* and MV *Pole Star* have a helideck constructed above their own decks to allow helicopters to pick up personnel, fuel, water and building materials.

Another trend common to both lighthouse authorities that had started well before the 1970s was automation. In fact the process started in the 1920s – long before silicon chips were dreamt of. It was a policy dictated to a large extent by simple economics and advancing technology. To any lighthouse service the more keepers it employs the more it is going to cost to run that service, and every day's delay in relieving those men is a further expense. Both the Northern Lighthouse Board and Trinity House followed a policy in the last twenty years of the twentieth century to convert all of their isolated and remote stations, followed by all their shore stations, to automatic operation.

From a purely financial and cost-efficiency standpoint, it makes absolute sense to install equipment whose operation can be made automatic and monitor it from a shore station via VHF radio or telemetry. Trinity House and the NLB cannot be criticised for pursuing such a policy, yet a consequence of its adoption was, of course, the elimination of the need for any keepers to be relieved at all – only regular maintenance visits, which might only take a few hours every six months, being required at each station.

The first automatic lighthouses were powered by acetylene gas under pressure which illuminated a gas mantle and turned the revolving lantern at the same time. A regular programme of replenishment of the acetylene cylinders was still needed, but thanks to a clever device known as a 'sun valve' the supply of acetylene was cut off during hours of daylight making it function like a traditional lighthouse – but without the expense of keepers. Chicken Rock lighthouse was converted to automatic operation in this manner after the fire of 1960.

When the first edition of this book was published in 1983, the automation process of rock lighthouses had only just begun in earnest with the automation of the Eddystone in 1982. By the end of 1998 it was complete – there wasn't a single manned Trinity House shore station

or rock lighthouse. It was much the same picture with the NLB who had retired all of their keepers by March of the same year. All of this has been possible due to rapid technological advances over the past few decades.

Despite all this, as I have already said, in the age of the silicon chip we have yet to find an adequate replacement for a lighthouse. What the microchip has done though is to be a more than adequate replacement for the lighthouse keeper. Equipment controlled by microprocessors that requires maintenance only twice a year instead of twice a day doesn't need the constant supervision and attention of a keeper. The main reason for a keeper to be at a lighthouse has therefore disappeared. Of course, the keepers provided other 'unofficial' services such as giving assistance to the RNLI or HM Coastguard when vessels got into trouble, or keeping weather records for the Meteorological Office – services which although not in their 'official' job descriptions, all keepers did willingly. The thousands of yachtsmen who spend their weekends and holidays sailing around our coasts will perhaps feel that little bit less secure when they realise that the lighthouses they pass

Above: The Trinity House helicopter lifts off from Wolf Rock lighthouse after changing the keepers. *(MBB)*

no longer contain any human observers to keep an eye on their progress – just in case the unexpected should happen.

What of the future? We are undoubtedly in a new era for the British lighthouse service. An era that, even without keepers, sees our two lighthouse authorities continue to improve the service they give and reduce the costs of providing that service. Currently, top of the lighthouse authority efficiency agenda is the process of solarisation – the conversion of lighthouses to make use of environmentally friendly – and free – solar power, rather than the costly, difficult to transport, awkward to store and potentially hazardous diesel oil. Development of high efficiency metal halide bulbs has allowed the replacement of the gas mantle and traditional lighthouse 'bulb' by solar-electric systems which charge the internal batteries by generating an electric current directly from sunlight. In fact, not even direct sunlight is required – something that can be in short supply in the north of Scotland during winter. The solar panels can make use of diffused light through cloud cover, and the batteries are of a sufficient number and size to accumulate enough energy in the summer and autumn months to ensure it can be stored for winter operation. Trinity House is now the largest user of solar power in the United Kingdom.

Above: The Trinity House helicopter is sat on top of Longships lighthouse with its clam-shell doors open. Two of the keepers are busy lowering supplies through the trapdoor to the lantern gallery below. *(Author)*

Part of the automation process can often involve reducing the range of the navigation light or altering its character. Modern commercial shipping equipped with satellite navigation and radar doesn't need a lighthouse to flash a warning that extends for over 20 miles as it used to do. The amount of power required for the reduced light range of, typically, 18 miles has decreased in the last decade from over 100 watts to 35 watts by making use of the new generation of high-efficiency and low-energy lamps operating directly from a 24-volt battery supply. Supplementary power for monitoring purposes can also be supplied from small wind-powered generators – and wind is something the north of Scotland has plenty of during winter!

Recent additions to many of the rock lights in this volume have been banks of solar panels in an appropriate location, out of reach of the waves. Usually, on the towers, the only place for them is attached to the outside of the lantern balcony. Eddystone, Wolf Rock and Hanois lighthouses have impressive rings of panels circling their helideck supports. At the same time as the solar panels were being added, many of the towers were losing their fog warning devices. In January 2005, the three General Lighthouse Authorities of the UK and Ireland issued a consultation document following a joint review of Aids to Navigation of the coasts of the United Kingdom and Ireland. Part of this document reviewed the need for the provision of fog horns. The conclusion was that audible fog signals had a significantly reduced role in the modern marine environment, as a result of the widespread use of electronic position-finding aids and radar, and the adoption of enclosed bridges on many vessels. There are now no Scottish lighthouses with a fog horn – the last to lose its 'voice' was Skerryvore in October 2005.

Left: Looking up from the lantern gallery of the Smalls lighthouse at a keeper on the helideck lowering his belongings through the trapdoor. *(Author)*

So, the particular era of achievement that produced the rock lighthouse, with its own special ethos, has gone forever. Part of the fascination and appeal of these structures was knowing that they were inhabited by men whose job was to be locked in the forefront of the

Right: The electronic sophistication of the 'Met' room in Bell Rock lighthouse with Principal Keeper John Boath (left) and Assistant Keeper Ernie England. The room is packed with all manner of warning, recording and communications equipment – as well as an exercise bike ! Note that there is still a telescope hanging from the ceiling.
(Jim Bain)

Right: A 1980s view of the 'modernised' Eddystone kitchen on the 6th floor.
(ALK archives)

struggle between man and the elements. A lifeless stone pillar, its insides packed with sophisticated electronic gear and monitoring devices, its exterior bristling with antennae, beacons and solar panels does not hold the same fascination. Perhaps the saddest result of automation is that without men living in the same quarters, using the same stairs, attending the same lantern, and looking across the same waters as their forerunners did over decades past, the stories of the raising of granite pillars out of foaming seas with all the human dramas that have taken place during and since that time, may slip quietly into oblivion.

A lighthouse without keepers becomes simply an impassive stone monument. When the interest of the local community ceases to focus upon that monument, as it surely has when the regular change of keepers ceases, when families cease to keep an extra special eye on the weather on relief days, or when the uncertainty of knowing whether their loved ones will be home for Christmas is gone, then its very existence very quickly fades from the memory, particularly if its remoteness also places it out of sight. It was the keepers who related the stories of the lighthouse, tales which had been passed down from previous keepers, tales of shipwreck and storm, so that such stories were perpetuated and kept within the memory – but only so long as the keepers remained within the tower. Now the day has come when there is no such person as a lighthouse keeper I fear the lighthouses themselves may become just maritime curiosities to future generations. It is partly because of such a fear that this volume has been produced.

Right: All the comforts of home? Quality fixtures and fittings inside the kitchen of Wolf Rock lighthouse in October 1993 after it was automated.
(Andreas Koellner & Claudia Beurer)

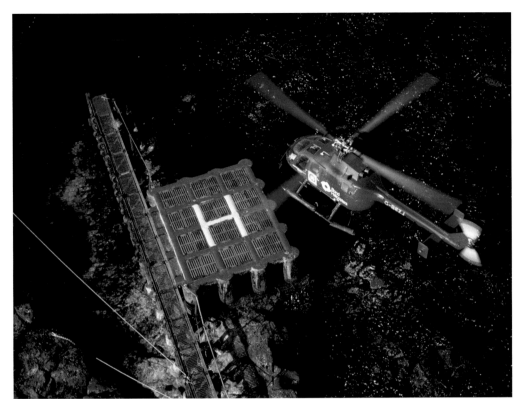

Left: The specially built helipad on the reef at Bell Rock meant keepers could be changed and equipment delivered without undue delay – providing it wasn't foggy or high tide of course. The NLB helicopter drops onto the helipad in ideal conditions.
(Jim Bain)

Left: The keepers that have just been delivered to the Bell Rock lighthouse wave farewell to the NLB helicopter as it departs. They will be on duty here for the next month before it returns to take them ashore.
(Author)

BIBLIOGRAPHY

Books

Adams, W.H. Davenport, *The Story of our Lighthouses and Lightships* (Nelson, 1891).

Allardyce, Keith, & Hood, Evelyn M, *At Scotland's Edge* (Harper Collins 1996)

Allardyce, Keith, *Scotland's Edge Revisited* (Harper Collins 1998)

Armstrong, Warren, *The True Book about Lighthouses and Lightships* (Muller, 1958).

Armstrong, Warren, *White for Danger* (Elek, 1963).

Beaver, Patrick, *A History of Lighthouses* (Peter Davies, 1971).

Chadwick, L., *Lighthouses and Lightships* (Dobson, 1971).

Elder, Michael, *For Those in Peril – The Story of The Lifeboat Service* (John Murray, 1963).

Fisher, James, *Rockall* (The Country Book Club, 1957)

Grosvenor, J., *Trinity House* (Staples Press, 1959).

Guichard, Jean & Trethewey, Ken, *North Atlantic Lighthouses* (Flammarion, 2002)

Hague, Douglas B. & Christie, Rosemary, *Lighthouses – Their Architecture, History and Archaeology* (Gower Press, 1975).

Hague, Douglas B., *Lighthouses of Wales* (Royal Commission on the Ancient and Historical Monuments of Wales, 1994)

Jackson, Derrick, *Lighthouses of England and Wales* (David & Charles, 1975).

Jerrome, E.G., Lighthouses, *Lightships and Buoys* (Basil Blackwell).

Lane, A.J., *It Was Fun While It Lasted* (Whittles Publishing, 1998)

Majdalany, Fred, *The Red Rocks of Eddystone* (Longmans, 1959).

Mair, Craig, *A Star For Seamen* (John Murray, 1978).

Mariotti, Annamaria Lilla, *The World's Greatest Lighthouses* (White Star Publishers, 2005)

Naish, John, *Seamarks – Their History and Development* (Stanford Maritime, 1985)

Noall, Cyril, *Cornish Lights and Shipwrecks* (Bradford Barton, 1968).

Ortzen, Len, *Famous Lifeboat Rescues.*

Plisson, Philip & Guillaume, *Lighthouses of the Atlantic* (Cassell & Co, 2001)

Smiles, Samuel, *Lives of the Engineers Vols 1 – 2* (London, 1861)

Stevenson, D. Alan, *The World's Lighthouses Before 1820* (Oxford University Press, 1959).

Sutton-Jones, Kenneth, *Pharos* (Michael Russell Publishing Ltd, 1985)

Williams, Archibald, *A Book Of The Sea* (Thomas Nelson & Sons, c1915)

Woodman, Richard and Wilson, Jane, *The Lighthouses of Trinity House* (Thomas Reed, 2002)

Zemke, Friedrich-Karl, *Leuchttürme der Welt Vol 1* (Koehlers, 1992)

Articles, reports and extracts

Beshaw, E.G. *Eddystone Lighthouse – Automation of a Rock Tower* (Flash, the magazine of the Trinity House Service, 1983).

Douglass, James Nicholas, *The Wolf Rock Lighthouse* (Proc. Inst. Civil Eng., March 1 1870).

Douglass, William Tregarthen, *The New Eddystone Lighthouse* (Proc. Inst. Civil Eng., November 27 1883).

Douglass, William Tregarthen, *The Bishop Rock Lighthouses* (Proc. Inst. Civil Eng., February 23 1892).

Hyslop, PH., *Repair of Fire Damage at Chicken Rock Lighthouse* (IALA Bulletin No 23, October 1964).

Hyslop, P.H., & Whitley Moran, T., *The Repair of Fire Damage at Skerryvore Lighthouse* (Inst. Civil Engineers booklet 1957).

Robertson, WA., *Manx Lighthouses and the Northern Lighthouse Board* (Journal of the Manx Museum, Vol VII, No 87, 1971).

Smeaton, John, *Narrative of the Building of the Eddystone Lighthouse* (1791).

Stevenson, Alan, *Account of the Skerryvore Lighthouse, with notes on Lighthouse Illumination* (1848).

Stevenson, David Alan, *The Dhu Heartach Lighthouse* (Proc. Inst. Civil Eng., April 25 1876).

Stevenson, Robert, *The Bell Rock Lighthouse* (1824).

Stevenson, Robert Louis, *The New Lighthouse on the Dhu Heartach Rock, Argyllshire* (The Silverado Museum, California, 1995)

Repairing the Chickens Rock Tower – uncredited – (Air World Vol 15, No 2 September/October 1962).

APPENDIX

A DETAILED DESCRIPTION OF MAJOR BRITISH ROCK LIGHTHOUSES

Notes

Lighthouse	The official name of the lighthouse. My addition of I, II, III etc indicates a succession of structures on the same site. Each one is detailed chronologically.
Latitude & longitude	The exact position of the lighthouse.
Geographical position	A brief description as to the location of the latitude and longitude position.
Date of construction	The period of construction of the lighthouse, as far as is known.
Description of structure	What the lighthouse looks like.
Designed by	The official 'designer(s)' of the structure. These are sometimes different to the person(s) who constructed it.
Constructed by	The person(s) who built the structure.
Light first exhibited	As accurately as research will allow.
Height of light above MHW	The height of the centre of the light source (in feet and metres) above the Mean High Water mark.
Height of tower	The height of the structure (in feet and metres) from the lowest course of masonry or rock surface to the top of the structure – usually measured to the top of the lantern room roof or weather vane etc. The accuracy of this, and the previous measurement, in the more historic structures is often difficult to verify and should be regarded as a guide only.
Light character	This is the present character of the lighthouse. It is highly likely that this will have changed over the course of the lifetime of the structure as improvements in illuminants, better optics, electrification and automation have all had their effect on the light character. For reasons of space and difficulty in accurate research, I have not tried to detail these changes. My abbreviations are as follows: W = White, G = Green, R = Red, Gp = Group, Fl = Flash; Iso = Isophase (equal periods of light and dark), Occ = Occulting (a steady light with a total eclipse at regular intervals with the period of darkness always less than the period of light). Numbers in brackets = number of flashes. e.g. W Gp Fl (2) ev 15 secs = 2 white flashes every 15 seconds
Nominal range	The nominal range of the present light in nautical miles. My comments regarding changes in light character over the history of a lighthouse apply also to this figure.
Light source	The present light source, described as accurately as can be determined. My comments regarding changes in light character over the history of a lighthouse apply also to this entry.
Candle power	The power of the present structure, or estimation of the strength of the light source in historic structures.
Fog warning apparatus	A description and character of the present fog warning device, if fitted.
Watched/unwatched	Whether the station has permanent keepers living in/at the lighthouse.
Relief	How the keepers were relieved, or the date of automation for an unwatched lighthouse
Notes	Interesting or explanatory notes about the history of each station.

TRINITY HOUSE ROCK LIGHTHOUSES

Lighthouse	Eddystone I	Eddystone II	Eddystone III	Skerries I
Latitude & Longitude	50°11'N, 4°16'W	50°11'N, 4°16'W	50°11'N, 4°16'W	53°25'N, 4°36'W
Geographical position	English Channel, 9 mls S of Rame Head, Cornwall	English Channel, 9 mls S of Rame Head, Cornwall	English Channel, 9 mls S of Rame Head, Cornwall	Irish Sea, 7 mls NE of Holyhead, Anglesey
Date of construction	1696-98	1699	1699	1714-17
Description of structure	Elaborate and ornamental wood and stone tower	Taller/broader tower with increased ornamentation	Tapering circular tower. Wood and stone framework sheathed by oak planks	Stone building with circular tower on top. Separate light-keepers cottage
Designed by	Henry Winstanley	Henry Winstanley	John Rudyerd	William Trench
Constructed by	Henry Winstanley	Henry Winstanley	John Rudyerd	William Trench
Light first exhibited	14 November 1698	1699	28 July 1708	4 November 1717
Height of light above MHW	c35ft (11m)	c75ft (22m)	c78ft (24m)	c99ft (30m)
Height of tower	80ft (24m)	c120ft (37m)	92ft (28m)	36ft (12m)
Light character	Fixed white	Fixed white	Fixed white	Fixed
Nominal Range (sea miles)	c9	c10	c9	Not known
Light source	Tallow candles	Tallow candles	Tallow candles	Coal fire in open grate
Candle power	c67	c67	c67	Not known
Fog warning apparatus	None	None	None	None
Watched/unwatched	Watched	Watched	Watched	Watched
Relief	Boat → Plymouth	Boat → Plymouth	Boat → Plymouth	Boat → Holyhead
Notes	The first sea rock lighthouse on an exposed site in the world. Winter gales passed over it so dimensions increased	Destroyed by 'The Great Storm' of 26 November 1703. Winstanley perished inside his tower	Destroyed by a fire started in the lantern on 2 December 1755 (see Eddystone IV)	Considered to be one of the two worst lights in the UK in 1777. Frequently obscured by fog. Modified 1804 (see Skerries II)

Lighthouse	Casquets I	Flatholm I	Smalls I
Latitude & Longitude	49°43'N, 2°23'W	51°22'N, 3°07'W	51°43'N, 5°40'W
Geographical position	English Channel, 5 mls W of Alderney, Channel Islands	Bristol Channel, 6 mls S of Cardiff, S.Wales	St George's Channel, 21 mls W of St Ann's Head, Dyfed
Date of construction	1723-24	1737	1775-76
Description of structure	3 circular granite towers in a triangular pattern	Circular stone tower with coal grate on the top	Oak and cast iron piles supporting an octagonal wooden lantern room
Designed by	Thomas le Cocq	William Crispe	Henry Whiteside
Constructed by	Thomas le Cocq and William Norman	William Crispe	Henry Whiteside
Light first exhibited	30 October 1724	1 December 1737	1 September 1776
Height of light above MHW	c90ft (27m)	135ft (31m)	c69ft (21m)
Height of tower	c45ft (14m)	72ft (22m)	71ft (22m)
Light character	Fixed, fixed white	Fixed	Fixed white above green, then fixed white
Nominal Range (sea miles)	Not known	Not known	13
Light source	Coal fires in closed lanterns, then Argand lamps	Coal fire in open grate	27 oil lanterns
Candle power	Not known	Not known	Not known
Fog warning apparatus	None	None	None
Watched/unwatched	Watched	Watched	Watched
Relief	Boat → St Peter Port	Boat → Swansea	Boat → Solva
Notes	The 3 towers were named St Peter, St Thomas and Dungeon. All were intended to be at the same height above sea level. Argand lamps installed 25 November 1790 by Trinity House	Tower seriously damaged by a storm in December 1790. Many complaints about the poor light quality so the tower was raised and modernised (see Flatholm II)	The first rock light to have a characteristic 'oak tree' outline. Replaced by the present tower when wave action undermined its foundations (see Smalls II)

TRINITY HOUSE ROCK LIGHTHOUSES

Lighthouse	Longships I	Skerries II	Lundy I	Flatholm II
Latitude & Longitude	50°04'N, 5°45'W	53°25'N, 4°36'W	51°11'N, 4°42'W	51°22'N, 3°07'W
Geographical position	English/Bristol Channels, 1 mile W of Land's End, Cornwall	Irish Sea, 7 mls NE of Holyhead, Anglesey	Lundy Island, Bristol Channel, summit of island	Bristol Channel, 6 mls S of Cardiff, S.Wales
Date of construction	1791-95	1803-4	1819-20	1819-20
Description of structure	Circular granite tower, 3 stories high	Taller version of original tower, now white with red bands	Circular granite tower, light-keepers cottage attached	Taller version of original tower, now painted white
Designed by	Samuel Wyatt	James Walker	David Alexander	William Dickinson
Constructed by	Samuel Wyatt	James Walker	Joseph Nelson	William Dickinson
Light first exhibited	29 September 1795	23 September 1846	21 February 1820	1821
Height of light above MHW	79ft (24m)	119ft (36m)	538ft (164m), 508ft (134m)	163ft (50m)
Height of tower	52ft (16m)	75ft (23m)	96ft (24m)	99ft (30m)
Light character	Fixed white	W Gp Fl (2) ev 10 secs	Upper: Quick Fl W ev 60 secs Lower: Fixed White	WR Gp Fl (3) ev 10 secs sets
Nominal Range (sea miles)	14	22	c30	W 15, R 12
Light source	19 Argand lamps	1kW metal halide bulb	Probably oil lamps	100W bulb
Candle power (approx)	Not known	1,150,000	Not known	W13,000, R3,200
Fog warning apparatus	None	Electric (2) ev 60 secs	None	Discontinued
Watched/unwatched	Watched	Unwatched	Watched	Unwatched
Relief	Boat → Sennen Cove	Automatic operation since September 1987	Dwelling attached	Automatic operation since January 1989
Notes	Heavy seas frequently passed over the tower and obscured the light. Replaced by a taller structure (see Longships II)	The last private lighthouse to be acquired by Trinity House after many legal difficulties. Extensively modernised 1967 and 1986. Now monitored from T.H. depot, Harwich	First English lighthouse to have a quick flashing light. Often obscured by fog so abandoned in 1897 when 2 new lighthouses were built (see Lundy North and South)	Bought by Trinity House from private ownership in March 1823 for £15,838. Converted to solar power, April 1997. Now monitored from T.H. depot, Harwich

Lighthouse	Bardsey Island	Longstone	South Bishop	Coquet Island
Latitude & Longitude	52°45'N, 4°48'W	55°39'N, 1°37'W	51°51'N, 5°25'W	55°20'N, 1°32'W
Geographical position	Cardigan Bay, 2¾ mls SW of Lleyn Peninsula, Gwynedd	North Sea, 18 mls SE of Berwick, Northumberland	St George's Channel, 5 mls SW of St David's Head, Dyfed	North Sea, 4 mls SE of Alnmouth, Northumberland
Date of construction	1821	1825-26	1839	1840-41
Description of structure	Square stone tower, painted white with red bands	Circular granite tower, red with white band. Lightkeepers house attached	Circular stone tower, painted white	Square sandstone tower, painted white, turreted parapet
Designed by	Joseph Nelson	Joseph Nelson	James Walker	James Walker
Constructed by	Joseph Nelson	Joseph Nelson	James Walker	James Walker
Light first exhibited	1821	15 February 1826	1839	1 October 1841
Height of light above MHW	129ft (39m)	85ft (26m)	144ft (44m)	85ft (25m)
Height of tower	99ft (30m)	85ft (26m)	36ft (11m)	72ft (22m)
Light character	W Gp Fl (5) ev 15 secs	W Fl (1) ev 20 secs	W Fl (1) ev 5 secs	W+R Gp Fl (3) ev 30 secs
Range (sea miles)	26	24	16	W21, R16
Light source	400W metal halide bulb	1kW metal halide bulb	70W metal halide bulb	12x200W sealed beam lamps
Candle power (approx)	667,000	640,000	60,000	W155,000, R21,830
Fog warning apparatus	Horn Morse 'N' ev 45 secs	Horn (2) ev 60 secs	Horn (3) ev 45 secs	Horn (1) ev 30 secs
Watched/unwatched	Unwatched	Unwatched	Unwatched	Unwatched
Relief	Automatic operation since July 1987	Automatic operation since September 1990	Automatic operation since January 1984	Automatic operation since December 1990
Notes	One of the few British rock lights with a square tower. Now monitored from T.H. depot, Harwich	Home of Grace Darling when she rescued survivors of the Forfarshire wreck. Modernised 1952. Now monitored from T.H. depot, Harwich	Special bird perches erected on the lantern by the RSPB and Trinity House to reduce migrating bird fatalities. Solar panels fitted in 2000. Now monitored from T.H. depot, Harwich	Tower built on top of former monastery. Now monitored from T.H. depot, Harwich

TRINITY HOUSE ROCK LIGHTHOUSES

Lighthouse	Lundy North	Round Island	Bishop Rock III	Eddystone V	Royal Sovereign	Skokholm	Beachy Head	Lundy South
Latitude & Longitude	5°12'N, 4°41'W	49°59'N, 6°19'W	49°52'N, 6°27'W	50°11'N, 4°16'W	50°43'N, 0°26'E	51°42'N, 5°17'W	50°44'N, 0°15'E	51°10'N, 4°39'W
Geographical position	Bristol Channel, northernmost point of Lundy Island	English Channel, northernmost of Scilly Islands	English Channel, westernmost reef of Scilly Islands	English Channel, 9 mls S of Rame Head, Cornwall	English Channel, 8 miles SE of Beachy Head, E.Sussex	St George's Channel, 4 miles NW of St Ann's Head, Dyfed	English Channel, 3 miles SW of Eastbourne, E.Sussex	Bristol Channel, SE corner of Lundy Island
Date of construction	1896–97	1886–87	1883–87	1878–82	1967–71	1916	1900–2	1896–97
Description of structure	Circular stone tower painted white, lightkeepers house attached	Circular granite tower painted white, lightkeepers house attached	Circular granite tower with white helideck	Circular granite tower with red helideck	Steel/ferro concrete platform supported on a single leg, with a white lantern	Octagonal stone tower painted white on top of lightkeepers house	Circular granite tower, red and white bands, red lantern	Circular stone tower painted white, lightkeepers house attached
Designed by	Daniel Alexander	James Douglass	James Douglass	James Douglass	Christians & Nielsen	Thomas Matthews	Thomas Matthews	Daniel Alexander
Constructed by	Joseph Nelson	William Douglass	William Douglass	James & William Douglass	Not known		Messrs Bullivant & Co.	Joseph Nelson
Light first exhibited	1897	1897	25 October 1887	18 May 1882	1971	1916	1902	1897
Height of light above MHW	156ft (48m)	180ft (55m)	144ft (44m)	135ft (41m)	93ft (28m)	177ft (54m)	103ft (31m)	175ft (53m)
Height of tower	56ft (17m)	63ft (19m)	160ft (49m)	159ft (49m)	117ft (36m)	58ft (18m)	142ft (43m)	52ft (16m)
Light character	W Fl (I) ev 15 secs	W Fl (I) ev 10 secs	W Gp Fl (2) ev 15 secs	W Gp Fl (2) ev 10 secs	W Fl (I) ev 20 secs	W Gp Fl (2) ev 20 secs	W Gp Fl (2) ev 20 secs	W Fl (I) ev 5 secs
Nominal Range (sea miles)	17	24	24	17	12	25	20	15
Light source	14W tungsten bulb	1kW metal halide bulb	400W metal halide lamp	70W metal halide bulb	35W halogen bulb	400 kW metal halide bulb	400W metal halide bulb	35W tungsten halogen bulb
Candle power (approx)	11,740	340,000	600,000	199,000	3,500	1,000,000	635,000	11,100
Fog warning apparatus	Discontinued 1988	Horn (4) ev 60 secs	Horn Morse 'N' ev 90 secs	Horn (I) ev 30 secs	Horn (2) ev 30 secs	Discontinued 1983	Horn (I) ev 30 secs	Horn (I) ev 25 secs
Watched/unwatched	Unwatched	Unwatched	Unwatched	Unwatched	Unwatched	Unwatched	Unwatched	Unwatched
Relief	Automatic operation since 1971	Automatic operation since 1988	Automatic operation since 15 December 1992	Automatic operation since 18 May 1982	Automatic operation since 19 August 1994	Automatic operation since December 1983	Automatic operation since July 1983	Automatic operation since December 1994
Notes	Built to replace Lundy I. Modernised 1971 and automated. Solar powered light installed 1991. Now monitored from T.H. depot, Harwich.	Contained one of the largest lantern optics in Britain until modernised in 1967. Now monitored from T.H. depot, Harwich	Bishop Rock II encased with another layer of masonry and increased in height. Said to be the most exposed lighthouse in Britain. Now monitored from T.H. depot, Harwich	The most famous lighthouse in the world. Replaced Smeaton's tower which was re-erected on Plymouth Hoe. Solar panels fitted 1999. Now monitored from T.H. depot, Harwich	The most modern Trinity House 'rock' lighthouse. Replaced a lightvessel stationed on the shoal since 1875. Now monitored from T.H. depot, Harwich	Extensively modernised and electricity installed 1963. Now monitored from T.H. depot, Harwich	Replaced original tower of 1828 on top of Beachy Head which was frequently obscured by fog. Now monitored from T.H. depot, Harwich	Built to replace Lundy I. Solar power installed 1994. Now monitored from T.H. depot, Harwich

TRINITY HOUSE ROCK LIGHTHOUSES

Lighthouse	Bishop Rock I	Casquets II	Bishop Rock II	Needles	Smalls II	Les Hanois	Wolf Rock	Longships II
Latitude & Longitude	49°52'N, 6°27'W	49°43'N, 2°23'W	49°52'N, 6°27'W	50°40'N, 1°35'W	51°43'N, 5°40'W	49°26'N, 2°42'W	49°57'N, 5°48'W	50°04'N, 5°45'W
Geographical position	English Channel, western-most of Scilly Islands	English Channel, 5 mls W of Alderney, Channel Islands	English Channel, western-most reef of Scilly Islands	English Channel, western extremity of Isle of Wight	St George's Channel, 21 mls W of St Ann's Head, Dyfed	English Channel, reef SW of Guernsey, Channel Islands	English Channel, 8 miles SW of Land's End, Cornwall	English Channel 1 mile W Land's End, Cornwall
Date of construction	1847–50	1854	1852–58	1859	1859–61	1861–62	1862–70	1870–73
Description of structure	Cast iron legs supporting a circular cast iron lantern room	Original towers increased in height and painted white with red stripes	Circular granite tower	Circular granite tower with red band and white helideck	Circular granite tower with helideck, and lantern	Circular granite tower with red helideck	Circular granite tower with grey helideck and granite landing platform	Circular granite tower with white helideck
Designed by	James Walker	Not known	James Walker	James Walker	James Walker	Nicholas Douglass	James Walker	William Douglass
Constructed by	Nicholas & James Douglass	Not known	James & Nicholas Douglass	James Walker	James & Nicholas Douglass	Nicholas Douglass	James, Nicholas & William Douglass	Michael Beazeley
Light first exhibited	Never operational	1854	1 September 1858	1859	7 August 1861	1862	1 January 1870	December 1873
Height of light above MHW	Never installed	120ft (37m)	110ft (34m)	80ft (24m)	W126ft (36m), R107ft (33m)	100ft (30m)	110ft (34m)	114ft (35m)
Height of tower	120ft (37m)	75ft (23m)	147ft (45m)	101ft (31m)	133ft (41m)	107ft (33m)	134ft (41m)	114ft (35m)
Light character	Never operational	W Gp Fl (5) ev 30 secs	Fixed white	WRG Gp Occ (2) ev 20 secs	W Gp Fl (3) ev 15 secs + Fixed R	W Gp Fl (2) ev 13 secs	W Fl (I) ev 15 secs	Gp Fl (2) ev 10 secs. W & R sectors
Nominal Range (sea miles)	30 (intended)	24	16	W17, R & G14	W25, R13	20	16	W15, R11
Light source	Not known	400W metal halide bulb	Argand lamps	1,500W bulb	400W metal halide bulb	35W metal halide bulb	1,500W bulb	3KW bulb
Candle power (approx)		1,079,000	Not known	W12,300 R3,950 G2,680	W1,000,600 R11,000	178,000	378,000	40,500
Fog warning apparatus	Bell suspended from the lantern gallery	Horn (2) ev 60 secs	Bell suspended from the lantern gallery	Horn (2) ev 30 secs	Horn (2) ev 60 secs	Horn (2) ev 60 secs	Horn (I) ev 30 secs	Horn (I) ev 10 secs
Watched/unwatched	Watched	Unwatched	Watched	Unwatched	Unwatched	Unwatched	Unwatched	Unwatched
Relief	Boat → St Mary's	Automatic operation since December 1990	Boat → St Mary's	Automatic operation since December 1994	Automatic operation since September 1987	Automatic operation since January 1996	Automatic operation since March 1988	Automatic operation since April 1988
Notes	Structure complete by the end of 1849 apart from the lantern. Destroyed by storm of 5 Feb 1850 before fully operational	Only the NW tower now functions as a lighthouse, others discontinued in 1877. E tower houses the fog horn. Now monitored from T.H. depot, Harwich	Spectacular storm damage led to tower being encased with another layer of masonry and raised (see Bishop Rock III)	Replaced an earlier tower of 1785 on cliff top which was frequently obscured by fog. Helideck added 1978. Now monitored from T.H. depot, Harwich	The most isolated Trinity House station. Replaced Henry Whiteside's unstable wooden light. Helideck added 1987, red & white stripes removed 1997. Now monitored from T.H. depot, Harwich	Electricity installed 1962, helideck erected 1979. The first UK rock lighthouse to be powered entirely by solar energy. Now monitored from T.H. depot, Harwich	The first lighthouse to have its keepers relieved by helicopter in November 1973. Converted to solar power 2003. Now monitored from T.H. depot, Harwich	Extensively modernised 1967. Helideck added 1979. Now monitored from T.H. depot, Harwich

NLB ROCK LIGHTHOUSES

Lighthouse	Isle of May I	Pladda	Inchkeith	Pentland Skerries I
Latitude & Longitude	56°11' N, 2°	55° 22'N, 5°07'W	56°02'N, 3°01'W	58° 41'N, 2°55'W
Geographical position	Firth of Forth, 12 miles N of Dunbar, Lothian	Firth of Clyde, 1¼ miles S of Isle of Arran, Strathclyde	Firth of Forth, 5 miles N of Leith, Lothian	Pentland Firth, 5 miles NE of Duncansby Head, Highland
Date of construction	1636	1789–90	1503–4	1794
Description of structure	Square parapetted stone tower, Coal grate on roof	Two circular stone towers, now painted white	Circular stone tower rising from square accommodation	Two circular stone towers 60ft apart
Designed by	Alexander Cunningham	Thomas Smith	Thomas Smith	Thomas Smith
Constructed by	Alexander Cunningham	George Shiells	Robert Stevenson	Robert Stevenson
Light first exhibited	1636	1 Oct 1790	14 September 1804	1 Oct 1794
Height of light above MHW	Not known	131ft (40m)	220ft (67m)	171ft (52m), 141ft (43m)
Height of tower	40ft (12m)	95ft (29m)	62ft (19m)	118ft (36m), 88ft (27m)
Light character	Fixed	W Gp Fl (3) ev 30	W Fl (1) ev 15 secs	Fixed white
Nominal Range (sea miles)	Not known	17	22	19, 18
Light source	Open coal gate	250W mercury vapour lamp	Electricity, sealed beam lamps	Oil lamps
Candle power (approx)	Not known	324,000	269,280	29,000
Fog warning apparatus	None	Discontinued	Discontinued	None
Watched/unwatched	Watched	Unwatched	Unwatched	Watched
Relief	Keepers in permanent residence	Automatic operation since 30 March 1990	Automatic operation since July 1986	Boat → St. Margaret's Hope, Orkney
Notes	Scotland's first lighthouse. Poor efficiency of the coal grate, so the lighthouse and island purchased by NLB and replaced by present tower		Robert Stevenson's first tower. Revolving light installed February 1815. Tower listed as being of architectural/historic interest. Now monitored from NLB HQ, Edinburgh	Lower tower built 1791 to distinguish it from other fixed lights in the same area. Now discontinued

Lighthouse	Bell Rock	Isle of May II	Calf of Man	Pentland Skerries II
Latitude & Longitude	56°26'N, 2°23'W	56°11'N, 2°33'W	54°03'N, 4°50'W	58°41'N, 2°55'W
Geographical position	North Sea, 12 miles SE of Arbroath, Tayside	Firth of Forth, 12 miles SE, of Dunbar, Lothian	Irish Sea, 1 mile SW of Isle of Man	Pentland Firth, 5 miles NE of Duncansby Head, Highland
Date of construction	1807–11	1814–16	1818–19	1833
Description of structure	Circular granite tower, now painted white	Square stone tower in gothic style rising from square accommodation	Two circular stone towers with lightkeepers' accommodation	Tallest of two original stone towers, now painted white
Designed by	Robert Stevenson	Robert Stevenson	Robert Stevenson	Robert Stevenson
Constructed by	Robert Stevenson	Robert Stevenson	Robert Stevenson	Robert Stevenson
Light first exhibited	1 February 1811	1 February 1816	1 February 1819	1833
Height of light above MHW	92ft (28m)	239ft (73m)	375ft (114m), 282ft (86m)	171ft (52m)
Height of tower	118ft (36m)	78ft (24m)	70ft (21m), 55ft (17m)	118ft (36m)
Light character	W Fl (1) ev 5 secs	Fixed white, now W Fl (1) ev 20 secs	Synchronised white revolving	W Gp Fl (1) ev 30 secs
Nominal Range (sea miles)	18	22	24, 22	25
Light source	35W lamp	250W mercury vapour lamps	Not known	250W metal halide lamp
Candle power (approx)	1,900,000	1,250,000	Not known	710,000
Fog warning apparatus	Discontinued 1988	Discontinued	None	Discontinued June 2005
Watched/unwatched	Unwatched	Unwatched	Watched	Unwatched
Relief	Automatic operation since 26 October 1988	Automatic operation since 31 March 1989	Boat → Port St Mary	Automatic operation since 31 March 1994
Notes	The oldest existing rock tower in Britain, Robert Stevenson's finest work. A severe fire in September 1987 delayed the automation programme. Now monitored from NLB HQ, Edinburgh	Replaced the original coal grate of 1636. Lighthouse building listed as being of architectural/historic interest. Now monitored from NLB HQ, Edinburgh	The synchronised beams were intended to mark Chicken Rock. Discontinued 1 January 1875 when Chicken Rock light erected	Rebuilt from tallest of two original optical towers. Tower listed as being of architectural/historic interest. Now monitored from NLB HQ, Edinburgh

NLB ROCK LIGHTHOUSES

Lighthouse	Muckle Flugga	Sanda	Skerryvore	Barra Head
Latitude & Longitude	60°51'N, 0°53'W	55°16'N, 5°35'W	56°19'N, 7°07'W	56°47'N, 7°39'W
Geographical position	North Sea, 1 mile N of North Uist, Shetland	Firth of Clyde, 2½ miles S of Mull of Kintyre	Atlantic Ocean, 12 miles SW of Tiree, Inner Hebrides	Atlantic Ocean, W end of Berneray Island, Outer Hebrides
Date of construction	1854–58	1850	1838–44	1830–33
Description of structure	Circular brick tower, painted white	Circular stone tower, painted white	Circular granite tower	Circular granite tower, painted white
Designed by	David & Thomas Stevenson	Alan Stevenson	Alan, and later Thomas Stevenson	Robert Stevenson
Constructed by	David & Thomas Stevenson, and Alan Brebner	Alan Stevenson	Alan Stevenson	Robert Stevenson
Light first exhibited	11 October 1854	1850	1 February 1844	15 October 1833
Height of light above MHW	217ft (66m)	164ft (50m)	151ft (46m)	682ft (208m)
Height of tower	64ft (20m)	48ft (15m)	138ft (42m)	59ft (18m)
Light character	W Gp Fl (2) ev 20 sets	W Fl (1) ev 10 secs	W Fl (1) ev 10 secs	W Fl (1) ev 15 secs
Nominal Range (sea miles)	22	15	23	18
Light source	250W metal halide lamp	250W mercury vapour lamp	250W metal halide lamp	Acetylene
Candle power (approx)	281,000	W61,000	1,000,000	162,000
Fog warning apparatus	None	Discontinued	Discontinued October 2005	None
Watched/unwatched	Unwatched	Unwatched	Unwatched	Unwatched
Relief	Automatic operation since 20 March 1995	Automatic operation since 26 March 1992	Automatic operation since 31 March 1994	Automatic operation since 23 October 1980
Notes	Temporary light built in 1854 until permanent tower operational in 1858. The most northern lighthouse in the UK. Now monitored from NLB HQ, Edinburgh	Solar panels and wind generators installed as part of the automation program. Now monitored from NLB HQ, Edinburgh	Scotland's most graceful tower and Alan Stevenson's greatest work. Lengthy automation process involved removing blue asbestos. Site of the last Scottish fog horn. Now monitored from NLB HQ, Edinburgh	The highest elevation of any British lighthouse. Now monitored from NLB HQ, Edinburgh

Lighthouse	Auskerry	Monach Isles	Fladda	Out Skerries
Latitude & Longitude	59°01'N, 2°34'W	57°32'N, 7°42'W	56°15'N, 5°41'W	60°25'N, 0°43'W
Geographical position	North Sea, 14 miles NE of Kirkwall, Orkney	Atlantic Ocean, 7 miles SW of North Uist, Outer Hebrides	Firth of Lorne, 15 miles SW of Oban, Strathclyde	North Sea, 20 miles NE of Lerwick, Shetland on Bound Skerry island
Date of construction	1865–66	1997	1860	1856–58
Description of structure	Circular brick tower, painted white	Originally 1864 circular red brick tower, now modern aluminium frame tower	Circular stone tower, painted white	Circular stone tower, painted white
Designed by	David & Thomas Stevenson	David & Thomas Stevenson	David & Thomas Stevenson	David & Thomas Stevenson
Constructed by	David & Thomas Stevenson	David & Thomas Stevenson	David & Thomas Stevenson	David & Thomas Stevenson
Light first exhibited	1 March 1866	1 February 1864–1942	1860	15 September 1858
Height of light above MHW	112ft (34m)	154ft (47m)	43ft (13m)	144ft (44m)
Height of tower	112ft (34m)	133ft (41m)	42ft (13m)	98ft (30m)
Light character	W Fl (1) ev 20 secs	W Fl (2) ev 15 secs	W R G Gp Fl (2) ev 9 secs	W R Fl (1) ev 20 secs
Range (sea miles)	18	10	W11, R9, G9	20
Light source	Acetylene	Was paraffin vapour burner	Acetylene	35W metal halide lamp
Candle power (approx)	67,000	173,000	8,250 & R 2,250	159,000
Fog warning apparatus	None	None	None	Discontinued
Watched/unwatched	Unwatched	Unwatched	Unwatched	Unwatched
Relief	Automatic operation since 1961	Automatic operation when built	Automatic operation since 1956	Automatic operation since 7 April 1972
Notes		Light discontinued 1942. A new 'minor' light established on the island 22 September 1997		Replaced a temporary light on Grunay island exhibited since 15 September 1854. Now monitored from NLB HQ, Edinburgh

NLB ROCK LIGHTHOUSES

Lighthouse	Dubh Artach	Chicken Rock	Fidra	Ailsa Craig
Latitude & Longitude	56°08'N, 6°38'W	54°02'N, 4°50'W	56°04'N, 2°46'W	55°15'N, 5°06'W
Geographical position	Atlantic Ocean, 14 miles SW of Iona, Inner Hebrides	Irish Sea, 3 miles S of Isle of Man	Firth of Forth, 16 miles NE of Leith, Lothian	Firth of Clyde, 10 miles W of Girvan, Dumfries & Galloway
Date of construction	1867–72	1869–74	1885	1882–86
Description of structure	Circular granite tower, red band	Circular granite tower	Circular brick tower, painted white	Circular stone tower, painted white
Designed by	David & Thomas Stevenson	David & Thomas Stevenson	Thomas & David A. Stevenson	Thomas & David A. Stevenson
Constructed by	David & Thomas Stevenson, and Alan Brebner	David & Thomas Stevenson	Thomas & David A. Stevenson	Thomas & David A. Stevenson
Light first exhibited	1 November 1872	1 January 1875	1885	15 June 1886
Height of light above MHW	145ft (44m)	125ft (38m)	113ft (34m)	59ft (18m)
Height of tower	125ft (38m)	144ft (44m)	56ft (17m)	36ft (11m)
Light character	W Gp Fl (2) ev 30 secs	W Fl (1) ev 5 secs	W Gp Fl (4) ev 30 secs	W Fl (1) ev 4 secs
Nominal Range (sea miles)	20	13	20	17
Light source	Metal halide lamp	Propane gas	Electricity 240V	Acetylene
Candle power (approx)	89,000	46,000	92,000	40,000
Fog warning apparatus	Discontinued	Discontinued	None	Discontinued
Watched/unwatched	Unwatched	Unwatched	Unwatched	Unwatched
Relief	Automatic operation since October 1971	Automatic operation since September 1962	Automatic operation since October 1970	Automatic operation since March 1990
Notes	Now monitored from NLB HQ, Edinburgh	Converted to automatic operation after disastrous fire of December 1960. Now monitored from NLB HQ, Edinburgh	First group flashing light in Scotland. Now monitored from NLB HQ, Edinburgh	Lighthouse tower listed as being of architectural/historical interest. Solar panels installed as part of automation process. Now monitored from NLB HQ, Edinburgh

NLB ROCK LIGHTHOUSES

Lighthouse	Bass Rock	Hyskeir	Copinsay	Calf of Man II
Latitude & Longitude	56°05'N, 2°38'W	56°58'N, 6°41'W	58°54'N, 2°40'W	54°03'N, 4°50'W
Geographical position	Firth of Forth, 3 miles NE of North Berwick, Lothian	Sea of the Hebrides, 33 mls W of Mallaig, Highland	North Sea, 2 miles SE of Point of Ayre, Orkney	Irish Sea, 1 mile SW of Isle of Man
Date of construction	1902	1904	1914–15	1968
Description of structure	Circular stone tower, painted white	Circular stone tower, painted white	Circular brick tower, painted white, fog horn tower adjacent	White octagonal tower on granite building
Designed by	David A. Stevenson	David A. Stevenson	David A. Stevenson	R.J. Mackay
Constructed by	David A. Stevenson	David A. Stevenson	David A. Stevenson	R.J. Mackay & P.H. Hyslop
Light first exhibited	1 November 1902	1904	8 November 1915	24 July 1965
Height of light above MHW	151ft (46m)	134ft (41m)	259ft (79m)	305ft (93m)
Height of tower	66ft (20m)	128ft (39m)	53ft (16m)	36ft (11m)
Light character	W Gp Fl (3) ev 20 secs	W Gp Fl (3) ev 30 secs	W Gp Fl (5) ev 30 secs	W Fl (1) ev 15 secs
Nominal Range (sea miles)	10	24	21	26
Light source	20W bifilament lamp	250W metal halide lamp	250W mercury vapour lamp	250W metal halide lamp
Candle power (approx)	1,100	788,000	180,000	1,000,000
Fog warning apparatus	Discontinued	Discontinued	Discontinued October 2001	Discontinued August 2005
Watched/unwatched	Unwatched	Unwatched	Unwatched	Unwatched
Relief	Automatic operation since 21 December 1988	Automatic operation since 31 March 1997	Automatic operation since 31 March 1991	Automatic operation since March 1995.
Notes	Reduced to minor light status since automation. Now monitored from NLB HQ, Edinburgh	Now monitored from NLB HQ, Edinburgh	Fog signal tower demolished 1985. Now monitored from NLB HQ, Edinburgh	Now monitored from NLB HQ, Edinburgh

NLB ROCK LIGHTHOUSES

Lighthouse	Stroma I	Sule Skerry	Stroma II	Flannan Isles
Latitude & Longitude	58°42'N, 3°07'W	59°05'N, 4°24'W	58°42'N, 3°07'W	58°17'N, 7°35'W
Geographical position	Pentland Firth, 5 miles NW of Duncansby Head, Highland	Atlantic Ocean, 37 miles NE of Cape Wrath, Highland	Pentland Firth, 5 miles NW of Duncansby Head, Highland	Atlantic Ocean, 18 miles W of Lewis, Outer Hebrides
Date of construction	c1889–90	1892–95	1896	1895–99
Description of structure	Not known	Circular stone tower, painted white	Circular stone tower painted white	Circular stone tower, painted white
Designed by	Not known	David A. Stevenson	David A. Stevenson	David A. Stevenson
Constructed by	Not known	John Aitken	David A. Stevenson	George Lawson
Light first exhibited	1890	1895	15 October 1896	7 December 1899
Height of light above MHW	Not known	125ft (34m)	105ft (32m)	331ft (101m)
Height of tower	Not known	85ft (27m)	76ft (23m)	75ft (23m)
Light character	W Fl (1) ev 60 secs	W Gp Fl (2) ev 15 secs	W Gp Fl (2) ev 20 secs	W Gp Fl (2) ev 30 secs
Range (sea miles)	Not known	21	26	20
Light source	Petroleum spirit burner	Acetylene	250W metal halide lamp	35W metal halide lamp
Candle power (approx)	Not known	65,000	1,100,000	100,000
Fog warning apparatus	Not known	None	Discontinued June 2005	None
Watched/unwatched	Watched	Unwatched	Unwatched	Unwatched
Relief	Not known	Automatic operation since December 1982	Automatic operation since 21 March 1996	Automatic operation since 28 September 1971
Notes	Very little known about this lighthouse. Only operational for 6 years	Most isolated British lighthouse. Tower listed as being of architectural/historical interest. Now monitored from the NLB HQ, Edinburgh	Lens from Su e Skerry lighthouse installed during automation. Now monitored from the NLB HQ, Edinburgh	Site of Flannan Isles mystery in 1900. Converted to solar power 1999

NLB ROCK LIGHTHOUSES

Lighthouse	Ve Skerries	North Rona	Haskeir	Rockall
Latitude & Longitude	60°22'N, 1°49'W	59°17'N, 5°49'W	57°42'N, 7°41'W	57°36'N, 13°41'W
Geographical position	Atlantic Ocean, 10 miles SW of Esha Ness lighthouse, Shetland Islands	Atlantic Ocean, 40 miles NW of Cape Wrath lighthouse, Highland	Atlantic Ocean, 8mls NW of North Uist, Outer Hebrides	Atlantic Ocean, 235 miles SW of Flannan Isles
Date of construction	May – September 1979	1983–4	1997	1972
Description of structure	Circular concrete tower, painted white	White square tower and 3 buildings	Cylindrical fibre glass tower painted white, complete with lantern room and gallery	Circular metal cylinder, painted red
Designed by	R.J. Mackay	R.J. Mackay	William Paterson	AGA Pharos Marine
Constructed by	R.J. Mackay & J.H.K Williamson	R.J. Mackay & J.H.K. Williamson	William Paterson	AGA Pharos Marine
Light first exhibited	27 September 1979	15 March 1984	1 October 1997	June 1972
Height of light above MHW	56ft (17m)	375ft (114m)	145ft (44m)	72ft (26m)
Height of tower	53ft (16m)	29ft (9m)	25ft (9m)	3ft (1m)
Light character	W Gp Fl (2) ev 20 secs	W Gp Fl (3) ev 20 secs	W Fl (1) ev 20 secs	W Fl (1) ev 15 secs
Range (sea miles)	11	24	23	8
Light source	Bi-form electric lamp	250W metal halide lamp	70W metal halide lamp	Various
Candle power (approx)	1,660	700,000	Not known	Not known
Fog warning apparatus	None	None	None	None
Watched/unwatched	Unwatched	Unwatched	Unwatched	Unwatched
Relief	Automatic operation when built.	Automatic operation when built	Automatic operation when built	Automatic operation when built
Notes	Of recent construction to aid navigation into Sullom Voe oil terminal from the west		The NLB's 200th light and its most modern 'rock' lighthouse. Powered by 63 solar panels and 6 wind generators	Not actually a NLB light – but placed here for convenience. The most isolated navigational aid in the world – and the most unreliable!

INDEX
Figures in *italic* represent illustrations

Above: Smeaton's stump on the Eddystone has been subjected to over a century of mountainous seas and winter storms which are beginning to take their toll on this historic structure. A constant assault by wave and wind has started to dismantle the blockwork – as these dramatic pictures show only too clearly. Not only can we now see the intricate jigsaw of blocks that Smeaton laid in 1758, but also the holes for the trenails and a couple of the square marble joggles. Slowly but surely the sea is starting to return that particular arm of the reef to how it was in 1696 before Winstanley had built the first Eddystone lighthouse.
(John Nicholls)